The What and How of Modelling Information and Knowledge

C. Maria Keet

The What and How of Modelling Information and Knowledge

From Mind Maps to Ontologies

 Springer

C. Maria Keet
Department of Computer Science
University of Cape Town
Cape Town, South Africa

ISBN 978-3-031-39694-6   ISBN 978-3-031-39695-3 (eBook)
https://doi.org/10.1007/978-3-031-39695-3

© The Editor(s) (if applicable) and The Author(s), under exclusive license to Springer Nature Switzerland AG 2023

This work is subject to copyright. All rights are solely and exclusively licensed by the Publisher, whether the whole or part of the material is concerned, specifically the rights of translation, reprinting, reuse of illustrations, recitation, broadcasting, reproduction on microfilms or in any other physical way, and transmission or information storage and retrieval, electronic adaptation, computer software, or by similar or dissimilar methodology now known or hereafter developed.

The use of general descriptive names, registered names, trademarks, service marks, etc. in this publication does not imply, even in the absence of a specific statement, that such names are exempt from the relevant protective laws and regulations and therefore free for general use.

The publisher, the authors, and the editors are safe to assume that the advice and information in this book are believed to be true and accurate at the date of publication. Neither the publisher nor the authors or the editors give a warranty, expressed or implied, with respect to the material contained herein or for any errors or omissions that may have been made. The publisher remains neutral with regard to jurisdictional claims in published maps and institutional affiliations.

Cover illustration: © Siarhei / Generated with AI / Stock.adobe.com

This Springer imprint is published by the registered company Springer Nature Switzerland AG
The registered company address is: Gewerbestrasse 11, 6330 Cham, Switzerland

Paper in this product is recyclable.

Preface

One may wonder why even bother with a book about modelling when there are the large language models with apps like ChatGPT that are claimed to be taking over the world by storm. Among others, they don't make you understand stuff. Modelling does. My own acquaintance with modelling goes back a while. Meandering through disciplines in my studies made me veer towards modelling in one way or another, without realising it at the time, without guidance—until I stumbled into computing, initially at a sysadmin conversion course and again at university. Computing is full of models and modelling and, consequently, there is a plethora of types of models as well as methods, techniques, tools, procedures, and methodologies to create them. But it seems as if we're keeping the goodies to ourselves. Information modelling is a recurring theme across disciplines and conceptual modelling is useful for a wide range of tasks. And so it seemed like a good plan to link up the different types of models and procedures and reveal their connections, overlap, and mutually beneficial aspects in an accessible, non-computing specialist, manner.

The main aim of this book is to introduce a group of models and modelling of information and knowledge comprehensibly. Such models and the processes for how to create them help to improve on one's skills to structure thoughts and ideas, to become more precise, to gain a deeper understanding of the matter being modelled, to improve analytical skills, and to assist with specific tasks where modelling helps, such as reading comprehension and summarisation of text. A secondary aim is to introduce several types of models of increasing versatility to step-wise broaden the readers' view on, and therewith knowledge of, models and the process of modelling that expand the modelling 'toolbox', all the while taking into account each model type's strengths and weaknesses rather than trumping only one of them.

This book covers five principal declarative modelling approaches to model information and knowledge for different, yet related, purposes. It starts with entry-level mind mapping, to proceed to biological models and diagrams, onward to conceptual data models in software development, and from there to ontologies in Artificial Intelligence and all the way to Ontology in philosophy. Each successive chapter about a type of model solves limitations of the preceding one and turns up the analytical skills a notch. These what-and-how for each type of model is followed by an integrative chapter that ties them together, comparing their strengths and key characteristics, ethics in modelling, and how to design a modelling language. In so doing, we'll address key questions such as: what type of models are there? How do you build one? What can you do with a model? Which type of model is best for what purpose? Why do all that modelling?

A typical model for each type of model is illustrated with the subject domain of dance, facilitating comparison of characteristic emphases across the models. Each chapter includes additional examples of models drawn from multiple subject domains, which are as varied as fermentation, labour migration, plankton, and peace. You'll discover something new about cladograms and the Gene Ontology, and pick up facts about lyrebirds and lemonade through the design of different types of models.

The conceptual modelling topics easily could have been granted a very theoretical treatise. They weren't. The advanced types of models we'll cover each have at least one textbook describing the 'what' of them in greater detail and you can consult those if they pique your interest. Rather, here we focus on general principles of the what and how in an engaging way—compared to technical documentation at least—and a key aspect is to gradually start seeing a red thread linking the types of models and the procedures for how to design them. The intended audience for the book as a whole are professionals, academics, and students in disciplines where systematic information modelling and knowledge representation is much less common than in computing, such as in commerce, law, and humanities. Biologists, while familiar with modelling and creating diagrams, can learn about new techniques for modelling beyond their common biological models, and when used, obtain new in silico analysis methods with the models. That said, if a computer science student or a software developer needs a quick refresher on conceptual data models or a short solid overview of ontologies, then those two chapters will server them well. A researcher in modelling might find joy in seeing spelled out that techniques easily cross over to other disciplines and that there are common aspects across several types of models—and the opportunities for research that invites.

While at the start of writing the book, I thought I knew all about what I wanted to write about, the process of writing about that arc of modelling with ever increasing complexity and trade-offs made it clearer to me and, to the best of my knowledge, it's the first time that these dots are being connected. Then there were the details, some of which were things I had wondered about every now and then but needed a reason to find out. Among others, ascertaining whether mind maps are really beneficial, triple-checking that still no-one devised a systematic procedure to create biological models, and experimenting with the conceptual schema design procedure on another modelling language. And there were the joys of discovering there actually are rules for what constitutes a correct cladogram and why and of devising an initial model for the ontology of dance, to name but a few topics from my writing viewpoint. More types of models could have been included in this book, but the point I wanted to make was made with the types of models covered already. Finally, needless to say, any errors or misunderstandings of referenced sources in this book are mine.

Acknowledgements

Nothing happens in isolation. Productive research collaborations with colleagues and students also helped my further understanding of conceptual data modelling and ontologies. I'm grateful to my long-term research collaborators on these topics: Pablo Fillottrani, Zubeida Khan (Dawood), and Agnieszka Ławrynowicz, and also Alessandro Artale, Sonia Berman, Oliver Kutz, and David Toman. I also would like to thank the current and former students who've conducted their honours, masters, or doctorate research on topics that found their way into this book in one form or another: Kieren Davies, Kouthar Dollie, Chiadika Emeruem, Zubeida Khan, Nasubo Ongoma, Toky Raboanary, Tamindran Shunmugam, and Steve Wang.

Over the years, a few academics have been willing to take some from their busy schedule to explain and provide feedback on modelling, modelling languages, logics, and ontological analysis, notably in my early years as PhD student, for which I'm grateful. I'd like to mention in particular Alessandro Artale, Diego Calvanese, Nicola Guarino, Terry Halpin, Claudio Masolo, and Barry Smith. For biology, where modelling and diagramming is taught as if by osmosis, there would be multiple teachers who added their bit; the teaching excellence of Ad van Egeraat was most memorable.

I'm grateful for the feedback received on parts of drafts of the book, notably by Sonia Berman and Oliver Kutz, that enabled me to improve the explanations and correct errors. I also would like to thank the staff at Springer, and Ralf Gerstner in particular, for their help and support.

Last, but not least, there's life outside work over the time I was writing this book that made writing this book easier. They include instructors and fellow dancers at Confidance for the Ginga Ladies Flashmob 2022 and at Forever Dance Academy, and the writers' meet-up group for the fun writing games. Thanks also go to Adele Kannemeyer, for the idea of a theme throughout the book to illustrate each type of model in its own way. That it turned out to be the topic of dance is, perhaps, not all that surprising.

Cape Town, South Africa C. Maria Keet
June 2023

Contents

1 **Introduction: Why Modelling?** 1
   1.1 What Is a Model? 1
   1.2 Not All Models Are Equal 5
   1.3 The Plan 10
   References 12

2 **Mind Maps** 13
   2.1 What Are Mind Maps? 14
       2.1.1 On Determining Whether Mind Maps Are Beneficial 15
       2.1.2 What the Researchers Observed 17
   2.2 How to Create a Mind Map.. 18
       2.2.1 Targeting for the Right Size and Shape 20
   2.3 Limitations 22
   References 23

3 **Models and Diagrams in Biology** 25
   3.1 Reading a Diagram: Two Examples 26
       3.1.1 Fermenting Sugars into Alcohol, Acids, and Gas 27
       3.1.2 Who Eats Whom in the Ocean, at the Microscopic Level 29
   3.2 A Quest for Common Characteristics 32
       3.2.1 The Chemists and the Cladists 33
       3.2.2 One Diagramming Language for all Biological Models 37
   3.3 How to Create a Biological Diagram 39
   3.4 Limitations 43
   References 46

4 **Conceptual Data Models** 49
   4.1 What Is a Conceptual Data Model? 50
       4.1.1 The Beginnings—and Still Standing Strong: ER 51
       4.1.2 Conceptual Data Modelling Explosion 55
       4.1.3 On Turf Wars and Truces 60
   4.2 How to Develop a Conceptual Data Model 65
       4.2.1 A Conceptual Schema Design Procedure..................... 67
       4.2.2 Top-Down and Bottom-Up Approaches 70

	4.3	Limitations	75
	References		77

5 Ontologies and Similar Artefacts ... 81
- 5.1 What Is an Ontology, the Artefact? ... 83
  - 5.1.1 Syntax and Semantics ... 83
  - 5.1.2 Automated Reasoning ... 88
  - 5.1.3 An Ontology Is More Than Just a Logical Theory ... 92
- 5.2 Success Stories of Using Ontologies ... 93
  - 5.2.1 Data Integration With the Gene Ontology ... 94
  - 5.2.2 Outperforming the Scientists and Engineers ... 98
  - 5.2.3 Automatic Question Generation and Marking with Ontologies ... 100
  - 5.2.4 Ontologies as the Panacea? ... 101
- 5.3 Methodologies for Developing Ontologies ... 102
  - 5.3.1 Bottom-Up Approaches to Ontology Development ... 103
  - 5.3.2 Top-Down Approaches to Ontology Development ... 105
  - 5.3.3 A Dance Ontology ... 107
- 5.4 Limitations ... 110
- References ... 111

6 Ontology—With a Capital O ... 115
- 6.1 The Greeks and Then Some ... 117
- 6.2 Examples: Parthood and Stuff ... 120
  - 6.2.1 Revisiting UML's Aggregation Association and GO's Part-of ... 120
  - 6.2.2 What's Lemonade, Really? ... 126
- 6.3 How to Do an Ontological Investigation ... 130
  - 6.3.1 A Tentative Procedure ... 131
  - 6.3.2 The Ontology of Dance ... 133
- 6.4 Limitations ... 137
- References ... 138

7 Fit For Purpose ... 141
- 7.1 A Beauty Contest ... 142
  - 7.1.1 A Feature-Based Comparison of the Types of Models ... 142
  - 7.1.2 Comparison by Example ... 148
- 7.2 Ethics and Modelling ... 155
  - 7.2.1 Professional Behaviour and Practices ... 155
  - 7.2.2 A Model's Features and Modelling Pitfalls ... 156
- 7.3 Design Your Own Modelling Language ... 161
- References ... 167

8 Go Forth and Model ... 169
- Reference ... 173

Index ... 175

About the Author

C. Maria Keet is an Associate Professor with the Department of Computer Science at the University of Cape Town, South Africa. Her research focuses on ontology engineering, conceptual data models, and natural language generation within the area of knowledge engineering, which has resulted in some 150 publications, including an award-winning textbook on ontology engineering and several best paper awards. She has been Principal Investigator and participant in several research projects funded by the South African National Research Foundation, the European Union, and Department of Science and Technology. Services to the community and outreach activities include program chairing of the International Conference on Knowledge Engineering and Knowledge Management, the Symposium on Conceptual Modeling Education, the Resources track of the International Semantic Web Conference, membership of the executive of the International Association for Ontology and its Applications and the EKAW steering committee, and editorial board membership, and volunteering for the Wikimedia Foundation. Besides her PhD in Computer Science (2008) from the Free University of Bozen-Bolzano, Italy, following a BSc (honours) in Computing and IT (2004) from the Open University UK, she also holds an MSc in Food Science free specialisation (microbiology) (1998) from Wageningen University, the Netherlands, and an MA in Peace and Development Studies (2003) from the University of Limerick, Ireland. This is her third book.

Acronyms

3D	Three-Dimensional
ACE	Angiotensin-Converting Enzyme
AI	Artificial Intelligence
BFO	Basic Formal Ontology
CAD/CAM	Computer-Aided Design/Computer-Aided Modeling
CSDP	Conceptual Schema Design Procedure
DOLCE	Descriptive Ontology for Linguistic and Cognitive Engineering
DWI	Dirty War Index
ER	Entity-Relationship [model/modelling/diagram/language]
EER	Extended Entity-Relationship [model/modelling/diagram/language]
FaCIL	Framework for semantiC Interoperability of conceptual data modelling Languages
FCO-IM	Fully Communication-Oriented Information Modeling
GEM	General Extensional Mereology
GIS	Geographic Information Systems
GPT	Generative Pre-trained Transformer
HCI	Human-Computer Interaction
HTML	HyperText Markup Language
ICD10	International Classification of Diseases version 10
IT	Information Technology
JSON	JavaScript Object Notation
NIAM	Nijssen's/Natural Information Analysis Methodology
OBO	Open Biological Ontologies
OMG	Object Management Group
ORM	Object-Role Modeling
OWL	Web Ontology Language
RNA	Ribonucleic Acid
SBVR	Semantics of Business Vocabulary and Business Rules
SNOMED CT	SNOMED Clinical Terms
STEM	Science Technology Engineering and Mathematics

TREND	Temporal information Representation in Entity-Relationship Diagrams
UML	Unified Modeling Language
XML	eXtensible Markup Language

Introduction: Why Modelling? 1

Real models don't go with the trend, they set the trend.
— Amit Kalantri, Wealth of Words

Modelling. If you were to ask school children what it is, they'll likely mention a career on the catwalk for the select few men and women that meet a certain definition of beauty. Not the sort of modelling that anyone can learn. This book is about a different type of modelling, an activity that is not an end in itself merely to creativity, but one that has a model as concrete output emanating from the act of modelling. I hope you, the reader, will gain an appreciation of the intricacies of at least some of those types of models and modelling, and and maybe also a new or increased fascination with the underlying commonalities and complexities that those superficially harmless-looking things actually convey and entail.

1.1 What Is a Model?

What is a model in the sense we'll use it in this book? It's an abstraction, idealisation, approximation or simplification of reality or of what is intended or expected to become reality, or of our best understanding of reality we can attain. Very many types of models fit this description. They range from the very simple to the very complex, from the proverbial scribbles on the back of the envelope to accurate computation that may take several person-years to meticulously put together or demand costly supercomputers and a few months of patience to see the computing power make the model. If you've played with LEGO® or similar toys, you will have built a model of your dream house at least once. If you're into gadgets, you may consider buying a 3D printer, which prints objects based on models of the objects designed with a software application. And regardless of whether you agree with the climate change model predictions or the weather forecast model predictions, models they are.

One of those months-long electricity-guzzling variety of models is the 175 billion parameters-strong language model called GPT-3 that operates behind the interface of the ChatGPT app that went viral in late 2022, drawing a mixture of awe, fear, and ridicule. GPT-3 seems closer to daily life with its potential use in chat bots, translation systems, and student essay cheating and is good for many an article in the online news and opinion websites. A concrete example of a software-based model that was created manually over many years by many people, and still is being extended 25 years hence, is the Gene Ontology. It is used to advance science and is humbly indispensable in scientific research, from causative agents of widely occurring diseases such as malaria and COVID-19 to rare genetic diseases like adrenoleukodystrophy that messes with children's brains—albeit unsung in popular media.[1] The best the most popular search engine could produce on the first results page was a Wikipedia article on the Gene Ontology and after research and tools for it, on page 5, its Twitter account. The Gene Ontology should concern the public since it's pervasive in science and the tools scientists use, yet it never ever produced the fodder that the likes of ChatGPT produced. And perhaps that's why you should take note of it.

Setting aside the popularity contest, a feature that both types of model share, is that they try to be as faithful to 'what they know about the world' as attainable, with ChatGPT based on the data it was fed for training and the Gene Ontology based on scientific research outcomes. This also holds for financial models and climate models that predict a possible state in the future—anyway, they aim to be as realistic as possible to obtain the best outcome. Other models may be more economical with reality, and intentionally so, or focus only on the parts the modeller deems relevant. Maquettes, or physical models, of old sailing boats, aeroplanes in miniature, and apartment complexes to be built, are intentionally simplifications of the real thing, because the general impression will do just fine and they should look pretty. Language models, I can assure you, do not look pretty.

Undoubtedly, anyone will have seen, if not also built, models at some point in their life. Compared to building a toy house or a sandcastle, something like the diagrams in biology textbooks are rather disparate entities and may seem unconnected. Questions that come to mind are: *what type of models are there? How do you build one? What can you do with a model? Which type of model is best for what purpose? Why do all that modelling?*

The last question is the easiest to answer: to try to predict something, to structure the unstructured, or to figure out how at least a part of the world works without

---

[1] An informal introduction to GPT-3 and GPT-4 can be found at https://en.wikipedia.org/wiki/GPT-3 https://en.wikipedia.org/wiki/GPT-3 (last accessed on 29-5-2023). Regarding adrenoleukodystrophy and the gene ontology, see Schlüter et al. (2012) of the open access paper at https://www.ncbi.nlm.nih.gov/pmc/articles/PMC3277307/ (last accessed on 29-5-2023). We shall return to the Gene Ontology later.

being distracted by details that don't matter much. There are secondary benefits emanating from this short answer. The structuring helps getting better answers in an online search on the Web—e.g., Google's knowledge graph—and it helps learning new study material, such as creating a Mind Map as a summary of a piece of text. The models for prediction can give insight into different scenarios of probable futures, which then assist with choosing among alternatives in the here and now (e.g., "reduce CO_2 emissions, or else..."). Drawing up a model of the components and processes in the body and knowing how it works generates insight into what it means when some part malfunctions when we're ill. The 3D bioprinted human heart model was a thing of late 2020 already, which needed a realistic model to instruct the printer to actually print it.

On the question of how to build a model, an untrained person will resort to informed guesses. Even a trained person will first ask for certain parameters before deciding on a course of action. For instance, on whether they can repeatedly ask a domain expert if the draft model is to their liking and revise and refine according to the feedback, or if it's a one-time opportunity so that they have to probe deeper at that first meeting to tease out and take note of as much requests and desired features as they can. I've tried to recall my primary and secondary school learning material about building models, but I think we were just offered models as pretty diagrams, rather than also being tasked with building them. That was decades ago. The world has changed—even core concepts such as long division are taught differently—so maybe modelling too. Sadly, it doesn't seem to be the case, and even less so for teaching learners how to model. Where did that leave you on the spectrum of experience with modelling? If you were tasked with modelling everyday tasks, what will your models look like? For instance, if you had to draw a model about the topic of 'buying a house' or 'decorating a house', what would you do? You may choose whether that would be a task-oriented model or an object-oriented model. Or try to draw a diagram of what you know about COVID-19, or about holidays, pets, or kitchen appliances.

The vagary in the description of that drawing task makes it harder to do: without an idea of the purpose of the model, it's not easy to know where or how to start, what sort of things are supposed to go in a model, what notation to use, and when to stop and why. Not to mention assessing whether the model you may have drawn is any good, and good for what. A model about kitchen appliances for a recipe book will be noticeably different from a model for the database of the kitchen appliances retailer who needs to keep track of the inventory and sales, and that model will yet again be unlike a model for the mechanic who fixes broken appliances. The type of model you would design for these scenarios is the same; only the content would vary. Not only that, it may well be that the method or technique of creating them is the same. Genericity in procedures irrespective of the particular instance, is something that engineers aim for, and have been successful in. We shall look at such model development procedures later in this book, where we'll take a structured approach to modelling that is based on research into the methods and techniques of developing models.

Fig. 1.1 A monk preparing to chop up the Tree of Porphyry, ca. 1503. It was in vain, for soon thereafter the tree of knowledge was revived, watered, grew, and was to be investigated with increasing frequency to this day and, according to forecasts of Gartner and the like, also with a bright future noticeably in IT

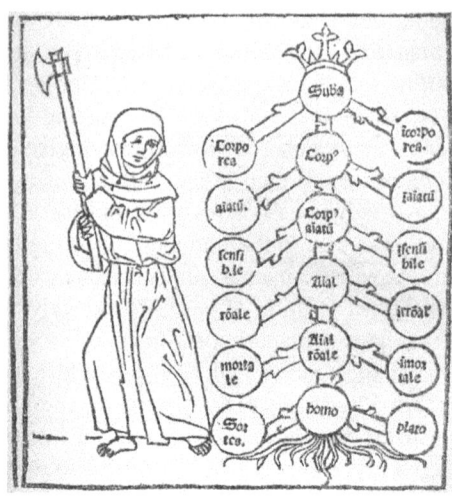

Before the structured approaches to designing models, however, came the models and modellers, not unlike chicken and traditional farmer before the entire process from egg to chicken was optimised in the twentieth century. Illustrious modellers of yore include the likes of Aristotle (third century BC), Plato (who taught Aristotle), and Porphyry (way later, third century CE). They tried to capture and structure the key things that exist, or anyhow were thought to exist according to their understanding of reality. Porphyry's tree modelled, among others, **Animal** as being either **Irrational** or **Rational**, with **Human** linked up to **Rational** and further divided into **This** and **That**, and at the end of the tree an individual rather than a genus or general property, which was **Plato** in that famous tree. Their flavour of interplay between modelling and scientific and philosophical inquiry fell out of favour, especially in the Middle Ages—the curious drawing by Augustinus de Ancona shown in Fig. 1.1 a case in point[2]—but eventually, logic and reason prevailed. As we shall see later in the book, that tree of knowledge has been revamped multiple times and more precise ways have been defined as foundations for the drawings. That increased precision comes with several benefits that we shall get to in due course as well. Be it for such lofty goals as the Greeks tried to aim for or, humbly, to organise one's thoughts in a systematic way, let's start bringing order to the myriad of models.

---

[2] De Ancona was a devout Italian theologian and philosopher. He was a contemporary of William of Ockham, who is known for 'Occam's razor', or the law of parsimony (to not use more entities than necessary).

1.2 Not All Models Are Equal

While 'model' and 'modelling' may sound specific when set against the many other happenings in life, they are teeming with respected huge differences, bun fights, and turf wars. How would I know? Modelling and models gained my interest over the second half of the 1990s at the dissertation stage of studying toward a degree in food science at Wageningen University in the Netherlands, where I specialised in microbiology. In near-absence of theoretical biology and bio-informatics at the time, however, I took a detour to, in 2003, start looking into it in earnest, first as an honours student in IT & Computing at the Open University UK, when I designed a database for bacteriocins and stumbled into awesome modelling problems to solve, and then as a young researcher at the Free University in Bozen-Bolzano in Italy studying toward a PhD in computer science. It's been a passion I get paid for ever since, conducting research into models and modelling for computing and many other domains, currently at the University of Cape Town in South Africa. And so, let's start nicely, with the respected differences, being four fundamentally distinctive notions of 'model', only one of which is the focus of this book.

A 'model' can be a physical model, like building your miniature dream house with LEGO® bricks or a sandcastle on the beach and it's common to build a maquette for a proposed fancy building. That may cost a lot of material and effort, especially when requirements change, and the physical model has to be redesigned. We can put models of 3-dimensional objects into software applications to play with them and iterate over different versions faster. The 3D modelling applications also enable the design of models for physical objects realistically, in that if the design doesn't obey physical laws, the software application will tell you it won't work before trying to actually build it. Such tools are known as CAD/CAM software, which is an abbreviation of Computer-Aided Design/Computer-Aided Modelling, which is heavily used in industrial engineering for cars and engines and the like.

The second group of models are mathematical models. There are so many of them, it has sub-fields of specialisation, such as climate models by climate scientists and epidemiological models by epidemiologists to predict the future for that particular topic. Those models have many mathematical formulae with variables such as air temperature, pressure, and wind magnitude.[3] A variable may be added or removed, or the value adjusted, a simulation run, and the outcome of the computations observed. For instance, one could change the value for the variable $Atmospheric\ CO_2\ level\ in\ ppmv$ to some other value and then recompute what the rise, or fall, of the global temperature would be. Or plug in a new variable into a COVID-19 epidemiological model, such as, say, the prevalence of a particular gene in the population that makes them more, $genPrevPos$, or less, $genPrevNeg$, susceptible to infection or to severe disease.[4] The actual variable names in those mathematical formulae are often abbreviations or mere one-letter symbols. To

---

[3] A brief explanation of the key ingredients in a climate model can be found at, e.g., https://climate.mit.edu/explainers/climate-models (last accessed on 29-5-2023).

illustrate, I looked up one that I'm more familiar with from my earlier studies: the model for bacterial or yeast growth in a closed system, like a kettle of hop and water in a brewery, There's a lag phase, an exponential phase, a stationary phase, and then death of the micro-organisms. For the exponential phase, there's a mean growth rate of the number of bacteria or yeast cells, which is a constant, k, that is calculated as:

$$k = \frac{log_{10}N_t - log_{10}N_0}{0.301 * t}$$

where N_t is the population at time t, N_0 is the initial population number, and t the time point such as the number of hours from the start. One can go on adding formulae to create a complete model of bacterial growth for a particular setting. Like whether it's all happening in a liquid with continuous nutrient delivery or continuous removal of waste product produced by those bacteria. The continuous removal option is necessary for the process of making Gouda cheese. When I wrote my exam for the introduction to microbiology course in early 1993, one of the exam questions required us to calculate how long it would take for the specified bacterial species to grow and multiply to drown us all in the sports hall where we were writing the exam, under the assumption of unlimited food supplies to the bacteria's liking. It was an entertaining course.

The third group of models are those from machine learning and deep learning. The ideas date back to 1763 with Bayes and 1676 with Leibniz, respectively, and more prominently since the 1950s. But it was not until the 1990s that we saw a surge in computational foundations, which became mainstream in the 2010s thanks to increases in massive amounts of data and computation, and it became popular in recent years by making it into the news almost daily.[5] The basic idea is to get your hands on lots of data and let an algorithm—i.e., a sequence of instructions, like those for baking a cake but then executed by the computer—loose on that data to attempt to find patterns in the data. The collection of the resulting patterns, which can be seen as variables in their own right, is called 'a model'. Then, when an unseen object is supplied to the algorithm, that is, an object that was not part of the dataset it created the model from, the unseen-before object will be assessed against that model to determine what's up with it. For instance, a model trained on cat photos is given a new photo with an object on it, and its output is then a probability of how likely that photo has a depiction of a cat. Or to automatically filter out résumés of female applicants as unsuitable because few female employees were

---

[4] The example is plausible thanks to recent genome-wide association studies (Kotsev et al. 2021), but it has not yet been included in epidemiological models, because more research is needed, both regarding the genes indicating increased susceptibility and what the prevalence of those genes would be the general population.

[5] There are a few timelines and the one of machine learning differs from the one of deep learning. One of the former can be found on Wikipedia at https://en.wikipedia.org/wiki/Timeline_of_machine_learning (last accessed on 29-5-2023) and the most recent one of the latter is presented by Schmidhuber (2022).

hired before. Or to translate "Sawubona" into "Good day" and translate swathes of text roughly with a degree of bias that can be traced back to unrepresentative data collection.[6] The quality of the models depends to some extent on the algorithms and to a considerable extent on the quantity and quality of the data that is used to 'train', i.e., create, the model. Past data may not be a good predictor for current times. Data that was collected for one purpose may not be suitable for another purpose. Data that happened to be idly lying around may not be a good sample of the population. For instance, tweets on Twitter are easily accessible and it's a popular source for Web research, but that's not an adequate random sampling for how the population in the world writes or what the population's sentiment is, nor even of the average internet user. There are many books about the techniques and books about challenges they face. This book is not one of them.

The fourth category may be described as *conceptual models*. They are models with concepts and the relationships between them, as a minimum, and possibly also with several constraints holding over that. The "concepts" may be entity types, object types, classes, unary predicates, or universals and the "relationships" may be called fact types, relations, or associations; the precise naming and their referents depend on the type of model. A popular family of languages for conceptual models in industry are those in the Unified Modelling Language (UML) family.[7] And there's the symbolic Artificial Intelligence (AI) part of AI, with ontologies and related logic-based theories and their weaker spin-offs with structured controlled vocabularies and knowledge graphs. It is this category of models that we're going to zoom into in the rest of the book. Two informal examples are shown in Fig. 1.2 in the domain of kitchen appliances: they each show a part of a model of espresso machines and their components. The one with the central concept and colourful lines is the low-hanging modelling fruit called mind maps; the other one is a UML class diagram with a few parts of the machines and types of parts. That UML diagram can be converted into program code or a database, or we can beef it up first to be more precise but then we're going to need a little bit of logic and a dash of philosophy. Those disciplines also deal with models, but use different terminology to describe it. Pedantically speaking, from a viewpoint of logics, the models are called theories rather than models (that refer to something else) and from a viewpoint of philosophy, the models are unlikely to contain 'concepts', but 'universals' or 'types' instead.

---

[6] Cat pictures is the running joke. It has been applied to a range of other objects, with greater or lesser accuracy; the review by Zhao et al. (2019) focuses on the deep learning and examples for wildlife conservation are summarised for a broader public by Tuia et al. (2022). On software for hiring processes, there are many models and issues in bias that hit the news over the past years. To the best of my knowledge, the Amazon case was one of the, if not the, first one where it came out in 2015 (Dastin 2018). A brief introduction on the bias in language models is presented by Yannic Kilcher here: https://www.youtube.com/watch?v=J7CrtblmMnU (last accessed on 29-5-2023). It has become a relatively widely known issue, which gained prominence with Google's firing of Timnit Gebru in 2020, but whose paper on bias in language models was published nonetheless (Bender et al. 2021).

[7] http://www.omg.org (last accessed on 29-5-2023).

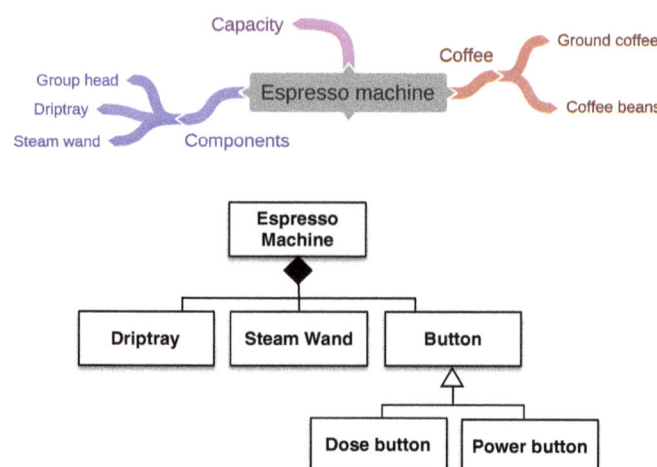

Fig. 1.2 Sample models of a selection of aspects of espresso machines

Historically, the name 'conceptual model' stuck, however, and it's not worth the energy to fight that name.

I'll readily admit that this fourth category of models is not the sexy new kid on the block these days, but they have snuck into a wide range of disparate tasks carried out by people in very many distinct roles. And they are making a come-back now that machine learning and deep learning are exhibiting limitations more pervasively. Put differently: they are ubiquitous even if you had not realised it and they are here to stay. They help accomplish tasks as diverse as resource flow modelling in companies and in ecosystems, analysis and design stages in software engineering, improving online search, discovering implicit knowledge for scientific discovery, and teaching reading comprehension in primary school. This entails that people with diverse specialisations actively engage with these types of models, such as, respectively: logistics managers, software engineers, anyone searching the Web, scientists, and learners and teachers. This list is not exhaustive by far. There are also the librarians with their categorisations and online retailers of the likes of Amazon with their hierarchies of product categories, to name but a few. Knowing of such models in some measure will help understanding those systems and therewith the chance to make them work better for you and to get more out of them. And/or to obtain a new tool in the toolbox of learning strategies, and a way to communicate with software engineers to get them to develop an application for you that meets the requirements better, and hopefully also extra insight into the nature of things.

They are bold claims, but reachable, albeit not with one single type of conceptual model. Besides huge differences between the four categories of models we have seen so far, there are also differences within the group of conceptual models. Some of them are respected differences; others have that distinct smell of, as warned in the introductory sentence, bun fights and turf wars in modelling. There aren't any

1.2 Not All Models Are Equal

between the 'big four', but within each type of modelling, or group of modelling languages, each with their methods and techniques to use for modelling, it's easy to ignite a bun fight among practitioners or experts. For instance, about which diagram notation is the best or which logic is best for the logic-based reconstruction of a given modelling language. The bun fights' topics are known among the experts, as are the various positions. There are both scientific arguments, of the variety with solid theoretical foundation or experimental evidence, and practical ones, like more software infrastructure. To appreciate those bun fights and, perhaps, join the fray, we'll have to dive into modelling first, which we'll see in the upcoming chapters, and especially in Chaps. 4 and 5. From my experience and vantage point, it doesn't look like those disagreements are going to be resolved any time soon or ever for a number of reasons. Live and let live, bicker over beer in a bar if you so fancy, and anyhow a good argument on those differences helps appreciating the pros and cons of the modelling languages better.

Then there are turf wars, small and large. As soon as anyone has obtained their PhD, that's their first substantial flag in the sand and turf to protect. Is the research community going to take up your solution or the one of another PhD student or researcher? Defend and promote your turf! They might become the stuff of small turf wars if the research topic is popular. The general public is blissfully unaware of them, as would most companies be. There are also medium-sized turf wars and big ones. The big ones amount to movements in one direction to substantial detriment of another. For instance, Artificial Intelligence (AI) conferences have been swamped with papers on 'the other' models—those data-driven ones—that crowd out, and perhaps even bully out, papers on the many other fields within AI that also work on models. Machine learning isn't even AI; they invaded AI turf.[8]

Fights and wars about matters that have more to do with the people involved than the scientific arguments, will be mostly ignored in this book. Scientific differences, on the other hand, can be very useful to better understand what the pros and cons are of one way of modelling over another or one diagram notation over another and therewith yield better arguments in the end. For instance, why conceptual models should not permit declaring attributes to have data types, like only declaring that persons have a height in centimetres but not whether that value should be recorded as a string (like '175cm') or as a integer ('175') or as float for accuracy ('175.00'), even though some types of conceptual models do allow a modeller to include that sort of information. But science and engineering are not democracies. What should you do then? Inform yourself and decide. This book aims to assist with that as well.

---

[8] Illustrative is the widely used classification of the Association for Computing Machinery, which is available at: https://dl.acm.org/ccs (last accessed on 29-5-2023). AI and machine learning are siblings under "computing methodologies", among other siblings, such as parallel and distributed computing methodologies and computer graphics.

1.3 The Plan

The main aim that I hope to achieve is that not only professionals but also the wider public will come to appreciate modelling and either learn to model or to do it better, which, in turn, assists with structuring thoughts and ideas more systematically, facilitate data analysis, and contribute to reading comprehension of texts. Why? Because the knowledge society needs a sharpening of analytical skills. Modelling is one of the tools in that toolbox. I hope that by the end of the book you, dear reader, will have gained some appreciation of the underlying theory as well as insights in practical aspects of designing models to experiment with creating a few yourself.

The book is structured accordingly: it ventures into a selection of the principal conceptual modelling approaches for distinct purposes, going from the entry-level mind mapping all the way up to philosophy—and back. This is also a key aspect that makes this book unique among books about modelling: there's no such book that travels along different types of models and modelling. There are trade books and university-level textbooks about a single type of model, but not taken together and compared against one another. While the chapters are ordered such that one type of model fixes issues of the one of the preceding chapter, this does not mean that the one that 'fixes' the issues with the previous type is necessarily more difficult. It has other features and strengths.

Specifically, Chaps. 2–6 introduce models and modelling with seemingly ever-increasing complexity—and possible use!—in a way that things are turned up a notch in each successive chapter. The one that comes after the other is the one that, purportedly or really, will solve some of or all the limitations of the one of the preceding chapter. Each one will be illustrated with a running example on the topic of dance. We'll start with mind maps, which you might vaguely remember from school. There's more to it, to the point there are still new insights being obtained by researchers looking into effective mind mapping. For the reader in Information Technology, computing, mathematics, or linguistics: they are trees with labelled edges and unlabelled nodes. We can do more and better, on how to display the contents, the details to show, the perspective to highlight, the preciseness of what we model, and the implications to extract from them. Turning up the precision a notch in the successive chapter, biological models may initially appear like mind maps, but there are additional types of elements, such as arrows, and implicit assumptions and notation conventions that become obvious once you know them. More can be said with biological models than with mind maps. How to do this for any subject domain, not just biology, requires a different strategy to modelling, which we'll see in Chap. 4: conceptual data modelling, the likes of which are used mainly in software engineering for any application in any subject domain. My gut feeling tells me that there is where one group of readers consider it generically usable enough to jump to Chap. 7, while another group may feel it's only getting to business then. The former group may nonetheless appreciate to stand on a mountain top to see there is another peak; the latter group will have gained appreciation that the first two climbs offered new vistas and will cease to call biological models 'cartoons'. Walking on,

1.3 The Plan

we'll arrive at ontologies in Chap. 5, which originally were intended to be models for a single subject domain to assist with integrating software systems, but over the years, ontologies have been used for many other tasks, computational and otherwise. For instance, as a way to get humans to agree or to help pinpoint where exactly the miscommunication or disagreement comes from, or to help categorise the results of major search engines on the Web. Taking a walk further down that road lands one in philosophy, in Chap. 6. We'll consider only that part of philosophy that has shown to be, or potentially is, relevant to conceptual modelling, which is a sub-field called analytic philosophy. There, one tries to reveal the nature of things and fundamental furniture of the universe, like Aristotle and Porphyry but then with the insights of today.

Be aware whilst going through this sequence that each type of conceptual model has its practitioners, champions, use in industry, and its researchers, and that the conceptual models that we see *en route* is a selection of all types of models proposed over the years. The spoiler alert, perhaps, is that neither of the types of modelling is perfect for every conceivable modelling task. We'll get to a comparison on a range of features in Chap. 7 to synthesise the respective pros and cons the types of models described in Chaps. 2–6. The strengths of each make them good for a particular subset of tasks.

Nowadays, a book on computing or Artificial Intelligence can't do without touching upon ethics in some way, and we'll do so as well in Chap. 7. It's possible to inject bias into models and mess up things, but the situation is not as bad as with the machine learning-based tools that try to predict the future based on historical or non-representative data. It was hard to find even a few real cases of breaches in ethics in modelling. Nonetheless, a few pointers will be given for things to watch out for. Another part of the synthesis and reflection is thinking outside the box whilst we're categorising things into boxes. Here, I'm referring to the notion of acquiring the capability of designing your own modelling language. Modelling as a tool normally comes with a ready-made toolbox. The tool and the toolbox didn't come out of nowhere, though, and who says you can't make them yourself? To not just accept the canon, but to take the machinery to boldly create the canon. The last section of Chap. 7 will be devoted to that. Finally, we'll wrap up in Chap. 8.

Let me close this introductory chapter with two recurring questions from my students and other time-constrained readers when they're faced with reading a book, and answer those. First: is it possible to skip sections or even whole chapters? You could try that, yes, but then the overall story won't be as much fun. It would be a bit like watching a webinar and the speaker breaks up every now and then: you may still get the gist of it if you're lucky, but might just be missing a crucial detail and be lost for the rest of the presentation. Unlike a live webinar, you can hit 'rewind', by reading at least the 'limitations' section of a chapter that you plan to skip or skim through; it's the last section of each chapter. The gist of the 'what' may still be graspable if you were to skip the 'how' sections; sponging information without intending to actively use it in building something is a feasible strategy. I did so with a few books I read for leisure.

The second not uncommon question is about whether chapters can be read out of order. Perhaps. Reading Chaps. 2–6 in reverse order might be feasible. I didn't verify whether is, but in theory it should: start with the complicated theory and simplify as you go along. But who does that? That only happens when one gives up on the complicated when it's too complicated, or when one does understand it but can't be bothered, or aren't up for killing a mosquito with a sledgehammer, or are willing to take certain downsides of simplification (together with the upsides). There will be fewer of those readers, I presume, and so we shall start with the colourful low-hanging fruit in modelling in the next chapter.

References

Bender EM, Gebru T, McMillan-Major A, Shmitchell S (2021) On the dangers of stochastic parrots: Can language models be too big? In: Proceedings of the 2021 ACM Conference on Fairness, Accountability, and Transparency. ACM, New York, pp 610–623

Dastin J (2018) Amazon scraps secret AI recruiting tool that showed bias against women. Reuters, October 11, 2018. https://www.reuters.com/article/us-amazon-com-jobs-automation-insight/amazon-scraps-secret-ai-recruiting-tool-that-showed-bias-against-women-idUSKCN1MK08G. Accessed 29 May 2023

Kotsev SV, Miteva D, Krayselska S, Shopova M, Pishmisheva-Peleva M, Stanilova SA, Velikova T (2021) Hypotheses and facts for genetic factors related to severe COVID-19. World J Virol 10(4):137–155

Schmidhuber J (2022) Annotated history of modern AI and deep learning. Technical Report Arxiv.org 2212.11279, KAUST AII, Swiss AI Lab IDSIA, USI. http://arxiv.org/abs/2212.11279

Tuia D, Kellenberger B, Beery S, Costelloe BR, Zuffi S, Risse B, Mathis A, Mathis MW, van Langevelde F, Burghardt T, Kays R, Klinck H, Wikelski M, Couzin ID, van Horn G, Crofoot MC, Stewart CV, Berger-Wolf T (2022) Object detection with deep learning: a review. Nat Commun 13(792):3212–3232

Zhao ZQ, Zheng P, Xu St, Wu X (2019) Object detection with deep learning: A review. IEEE Trans Neural Netw Learn Syst 30(11):3212–3232

Mind Maps

2

The mind map will change your life.
— Tony Buzan

Mind maps are *awesome!* With an exclamation mark indeed. You'll have to read Tony Buzan's books to catch a glimpse of just how amazing and exciting mind maps are. It's going to be a tad bit more down to earth here. Notwithstanding, mind maps still may be a revelation compared to a state of disorder—the great beyond *after* mind maps, which we'll see in the upcoming chapters, will then be double-plus exciting. Yet, when your daily tasks are not about dealing with models, or you find those models of the upcoming chapters a bridge too far, even sticking with mind maps will be useful. Mind maps for organising a meeting, planning a wedding, preparing you better for a job interview, setting up your own company, learning a language, and charting your life's vision and purpose—it's all possible. Now would be a good moment to mention an anecdote about the first time I heard about mind maps or when I made my first one, but I don't have such an anecdote. I can't recall when I first came across mind maps. I think mind maps may have passed the revue during my Bachelors in IT & Computing from the Open University UK around 2001, but not earlier, but I for sure stumbled over them reading scientific literature on knowledge acquisition to build a knowledge base several years later. I used a popular online search engine to double-check that mind mapping was what I thought it was and how it's dissimilar from concept maps. A South African friend's first reaction, who's about the same age as me but did not attend university, was "ahhh, yes, we did those in school!". Maybe it took a while before Buzan's marketing got translated into other languages and reached the village's school I attended, if it did so at all.

Mind maps do have a close affinity to education. Education researchers in particular, though also certain segments of industry, seem to think mind maps are the best things since sliced bread. They do serve a purpose and in the land of the blind, mind maps will make you the one-eyed king or queen. It might change your life—according to the main popularisers, it will. So, we're starting with that in this chapter. A seasoned modeller might look down on the mind mapper, snigger at their enthusiasm, and prefer to move on, but the odd nugget of information may give a refreshing view of mind maps after all and it can be used with those fancier modelling techniques. Could mind maps be just as good as those modelling techniques but with less effort? No, but there are benefits and we'll get to those. Let's first be more precise on what those mind maps really are.

2.1 What Are Mind Maps?

The question is easier than the answer. Tony Buzan, the chief evangelist of mind maps, said in his book that "A Mind Map is the ultimate organizational thinking tool—the Swiss army knife of the brain!", which does not reveal much on what it actually is. Nor do other phrases in the book with, e.g., "A mind map is a visual way to organise and learn information." as the key definition, nor does the process perspective rather than the object perspective, as "Mind mapping is a powerful technique to help you visually develop and organize ideas and information.". Another source has it both ways: it's a thing and a process.[1] These statements hold for those models that we'll see in the next two chapters as well, when the 'mind map' is substituted with those type of models. And yet, they are different beasts.

Let me try to formulate the answer to the question slightly more accurately: a mind map is an informal, structured, diagram whose creation assists in organising or learning related information by breaking down a central concept into its related component parts. It's a slightly longer sentence than the original tweet length formulation but does cover the duality of the object as outcome and the process of making one, with the actual tasks that take place in their development. Regarding that 'central concept' and component parts, if the central concept is, say, 'saving energy at school', then there will be several branches outward. For instance, a branch to how the school gets its electricity, which may have sub-branches, such as solar panels, burning waste, a windmill. Or how the school uses electricity for lighting, with sub-branches related to saving energy; e.g., electricity-efficient light bulbs, switching off lights during breaks and at night, and type of lighting fitting. Mind mapping is about the activity of coming up with those branches once you've chosen the central entity, and the outcome of the modelling exercise is a mind map that presents an overview of tasks and things that have something to do with the central entity. In colour. And with cartoons, if you like.

It's even possible to do this to mind mapping itself, that is, to create a mind map about mind mapping. This can go into several directions, depending on what's

---

[1] The first quote is by Buzan (2006); the second one from The Learning Fundamentals website (https://learningfundamentals.com.au/resources/); the third one is from https://www.mindmaps.com/what-is-mind-mapping; the last one is from mindmapping.com, where two successive paragraphs start with "A Mind Map is a. ...", at https://www.mindmapping.com/mind-map (last accessed on 29-5-2023).

2.1 What Are Mind Maps?

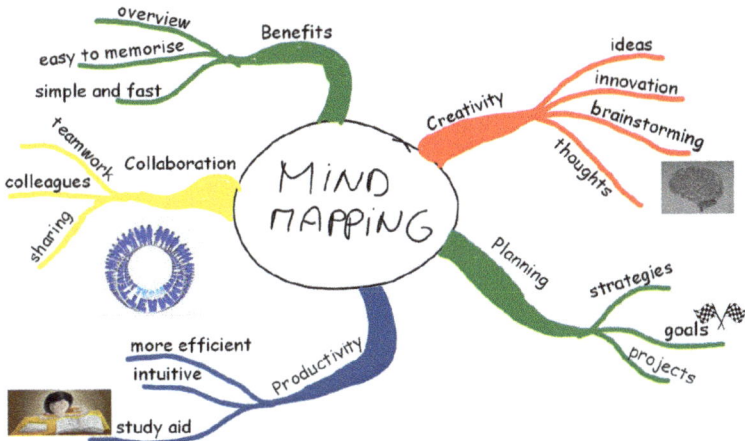

Fig. 2.1 A mind map about mind mapping, merging manual sketching with digital enhancements on a tablet device using the OpenBoard software, inspired by the manual drawing-like style of mind maps on popular websites such as mindmapping.com, mindmeister.com, and learningfundamentals.com.au, and Buzan's books

associated with it first. For instance, to focus on the benefits of mind mapping or on the activities involved in creating a mind map. A mind map about mind mapping is illustrated in Fig. 2.1, as one of the possibilities. Yours may look differently. To practice with the educational goals of mind maps, ignore the figure for now: when you're finished reading this chapter, create a mind map about your understanding of this chapter's contents, and only then have a look at Fig. 2.2, compare it with yours, and analyse why there are differences. Another mini-experiment to play with could be to draw another mind map about mind mapping after reading this book and compare it with the first mind map to see whether reading the book made you look at mind mapping differently. I hypothesise that will be the case.

2.1.1 On Determining Whether Mind Maps Are Beneficial

The before-and-after redrawing is merely a mini-experiment and the one anecdote it will result in, is not data and we shouldn't generalise from it. When I re-draw mind maps with the same central concept, they're never the same, but also that does not amount to data to say something meaningful about mind maps. Researchers have been looking into other aspects of mind maps. That is, not, mind maps in a 'before and after' intervention scenario but mind map 'as' intervention. A number of studies look at the use of mind mapping, notably on what people think of their use, and controlled experiments have been conducted especially in education research. Not all experiments are alike. Experiment participants' opinions can be collected in a survey, but opinion is not the same as usefulness or effectiveness. I enjoy the 2048 game app on my phone, but useful it is not. A grammar checker is really nice, but

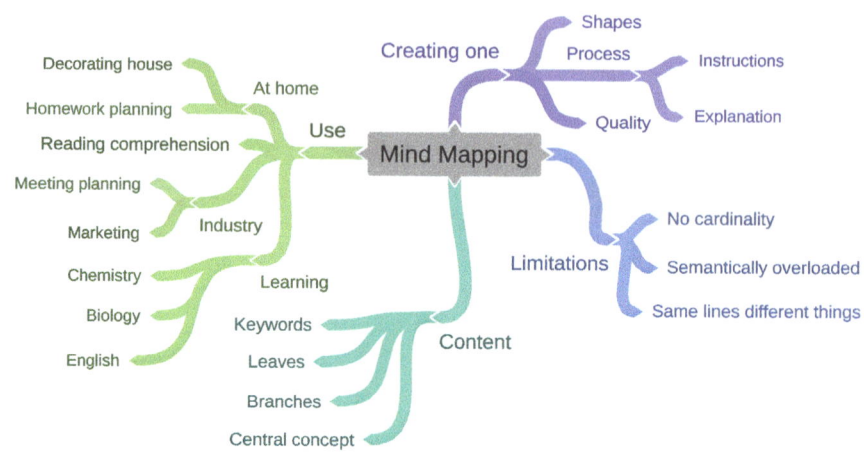

Fig. 2.2 A mind map about mind mapping topics of this chapter, made with online mind mapping software Coggle

effective in learning the grammar of a language, compared to studying the grammar, it is not. Maybe mind maps are useful for creating summaries, but not effective for obtaining a higher grade in the test. Or maybe they are effective but disliked.

Participants' opinions are relatively easy to obtain, as are their likes and dislikes, or self-perception of the usefulness: just ask. But for their use in learning, the key question any such experiment has to answer, is whether the learner learned better with mind maps or better without them. This still leaves a few variables that can be investigated. Among others: (1) how is 'better' measured? A higher grade in the test? Content retention six months later? Less time spent studying for the same grade? (2) What do the participants have to do with the mind maps? And (3) what subject are they supposed to be learning?

The second parameter consists of the choice between the teacher or experimenter giving the learners a mind map or have the learners make mind maps from the study material themselves. The act of creation, of structuring and summarising the content, is active learning, and works better for retention than passive sponging of content, and therefore the effects of mind maps should be larger when learners create them as compared to when they're given the mind map for rote learning. But maybe handwriting a summary is even better than drawing a mind map.

The third aspect, on what is being learned, can be as varied as a subject in school, such as chemistry, literature, biology, or processes at a company, including organising meetings or how to make the work environment greener. The amount of effect, if any, may vary by task or topic. All these variations already give 3 times 2 times n subjects/topics and processes to investigate to come to an exhaustive list and produce an answer.

We're nearly there. For experiments at schools, the straightforward set-up is to split up a class in two groups or have two classes by the same teacher for the same grade with the same content being taught, give one group the mind mapping as intervention, and the other group the regular teaching mode, test them on some dimension of the content—typically recall, understanding, or problem-solving—and compare the grades. If the grades are statistically significantly higher in the mind mapping group, then mind maps are beneficial for studying. But further dimensions can be added into the mix, such as the participants' background (e.g., culture, dyslexia) and level of study. The combinations of viable parameters and experiment set-ups are practically endless.

2.1.2 What the Researchers Observed

With all these variations in mind, is mind mapping shown to be beneficial? To a certain extent, in some configurations it has been shown that it is; in others, not. One can cherry-pick for confirmation bias, but that's not what is supposed to happen. Instead, a strategy out of the 'some do and some don't' is to conduct reviews and meta-analyses of the individual experiments to try to find any commonalities to finally obtain a conclusive answer either way. Ying Liu and co-authors from Beijing Normal University, China, were among the first who conducted a meta-analysis by pooling experimental results published between 1999 and 2013 to get larger numbers to arrive at more convincing statistical validity. They started out with 163 articles, but many articles had several issues and eventually at most 52 studies could be included. That analysis showed that effects were of the variety 'mostly beneficial to some extent, but not always'. Both software-based mind mapping and paper & pencil had a positive effect but not the three remaining unknown modes (those original papers didn't describe it). Also, the meta-analysis showed that mind maps had a comparatively large effect on learning English and the benefits were good for its use in the arts as well, but nil in geography and negative for medicine. The largest benefits were at primary school level, but much less so in middle school and college.[2] Very recently, in 2022, Yinghui Shi and co-authors from Central China Normal University in Wuhan, China, and Concordia University, Montreal, Canada, also conducted a meta-review, but they focussed on other parameters: whether there were any differences due to the pedagogical approach, like collaboratively working on mind maps or independently, how long the participants were subjected to mind mapping (more or less than one month), and the participation pattern, by which they meant student-generated versus student-generated and teacher or researcher-provided. They did not find any statistically significant difference for either of those parameters. That is, the experimental results did not support the hypothesis that mind maps are beneficial. It was statistically significant for positive effects in the Science Technology Engineering and Mathematics (STEM) cluster versus English and "other" subjects, which included medicine, history, and others. Also, Shi and

---

[2] See (Liu et al. 2014) for details.

co-authors found a larger positive effect in school compared to higher levels of education.[3]

No arguments were given in either meta-review why there are benefits for mainly STEM and lower levels of education. It's very tempting to speculate about it. I lean toward an assumption that the more one learns, the more complex things turn out to be, and to be able to convey that diagrammatically requires a graphical language of higher expressivity than just a root with colourful branches and cartoons. Why this is a plausible hypothesis will become evident in the next chapter, but a more fundamental 'why' has to be considered first and is, in fact, a more challenging one to answer: why does it work anywhere at all? A hunch for answer may be the platitude that 'one picture says a thousand words', but that is not a scientific answer. My search for an answer has not yielded one, and I tried to find one several times when authoring this book, because not finding the answer was mildly frustrating. If 'some structure is better than unstructured text' is the reason, then why not take a next step in structuring, to structure in greater detail and systematise the structuring? That is, take a logical next step from that premise. We will take those next steps in this book, both for mind maps and beyond. For now, since mind mapping is beneficial sometimes, the next step is to systematise how to create a mind map and to try to figure out what a good mind map looks like.

2.2 How to Create a Mind Map

The basic steps for creating a mind map are straightforward, once you know them. What makes it a less-than-trivial activity is the creative thinking or the reading comprehension. Let's assume you know enough of the central concept and want to go about drawing a mind map. Then the design steps can be as follows, which I put together from Buzan's book and I added a little to fill some gaps and make the implicit explicit.[4] The items are as follows and elaborated on afterwards:

1. Place your central concept at the centre of a blank page so that it allows for extensions in any direction equally. Optionally, add an image to that central concept.
2. Create your main branches, sub-branches, and twigs, writing a keyword on each of them, and, optionally, spice it up with cartoons here and there and a separate colour scheme for each main branch.

---

[3] See (Shi et al. 2022).

[4] The book: (Buzan 2006). It sort of describes the procedure by example on pp. 111–112, p. 126, and in the mind map for mind mapping on p. 130. Scientifically, there's more to 'design steps' and procedures, especially if one were to call the procedure a methodology: then it has to have been shown that procedure works better than no procedure (if it's the first procedure) or better than others in some way (if there are others to compare it with).

3. Connect the branches, be it at the root where your central concept is positioned or possibly also across branches.
4. Double-check you did not create a same branch or twig twice and that there are no 'orphans' that aren't connected to anything.

The 'blank page' in step 1 may be paper or the drawing canvas of a mind mapping application. Drawing on paper will be faster initially anyhow. The advantages of a good application are that it should be able to optimise the layout for you, any changes are easier erased or updated, and stock images should be easier to add compared to drawing them yourself. The images are said to improve recall of the mind map, as would the colour-coding help to remember the branches. It should easily offset the time it takes to find a mind mapping drawing tool you like. There are tools that are for free or cost money, are desktop applications or are online for easier collaboration, have differently coloured branches or not, lines of same thickness or not, and text on top of the branches, in the branches, or in ovals. There's no official standard for the notation, so the software developers choose for you, based on what they think you will like, what they like, and what they are able to program.

The key question is what to put on those branches and twigs. That is the non-trivial part of it all. Perhaps just the "keywords" mentioned in step 2? In a way, mostly yes, but at a deeper level: not quite. The things mentioned in mind maps at the places where it branches off are typically either *examples* of the preceding branch, *attributes* of it, or *parts* of the object or process. For instance, a mind map about doing homework may have a branch called "distractions". What could its branches be? Examples of distractions that prevent you from doing homework are social media, the TV, the Web, and perhaps the cat or dog or an energetic little brother or sister, too. The homework mind map may also have a branch called "doing exercises", which could be either workbook exercises or physical exercises to stretch regularly. The latter category may have branches with warm-up, running, weightlifting, and cool-down, which constitute parts of exercise sessions, rather than examples. Analysing the kinds of things at the branches has substantial potential to overcomplicate matters and is ripe and ready for research (we'll get to it in Chaps. 5 and 6), but you are likely doing it intuitively already anyway. And that is where it's supposed to stay with mind maps—to get the brainstorming going, rather than getting stuck in an analysis paralysis upon drawing the first branch.

To not just tell but also do, I've drawn a couple of mind maps and included a few in this book. I tried to create one about dance without analysing it whilst drawing, and the outcome is shown in Fig. 2.3. There's dance at the centre, it has branches with examples of various dance styles, related things that came to mind, like dressing up for the dancing, and movies and performances. Even a cursory analysis of it already makes me cringe because of the disparate topics having been put together haphazardly, but it looks nice and does give an impression of the dance domain. The activity did raise a new question: when is a mind map 'finished'? Or, if aiming for a pass: when is it sufficient and okay to stop?

Fig. 2.3 A mind map about dance

2.2.1 Targeting for the Right Size and Shape

There's no official definite end point of a mind map. There are no rules that say that you're allowed to have, say, only three levels from the central concept or at most 40 leaves at the ends of those branches. There's a physical limitation as to what can be put on an A4 sheet, so paper can be a blessing compared to near-limitless drawing in a software application. But do not let yourself be inhibited by a mere A4 sheet of paper: adding a new sheet and gluing or taping them together is very well doable. Practically, it's done when you think it's done. That circular specification is unsatisfactory and the researcher in me wondered about what the 'typical shape' of the average mind map would be and what an 'average mind map' amounts to. I set off to find out.

The first strategy was to search for scientific literature in the hope someone had asked the same question before and already figured out the answer, but that did not return anything. The next step was the proverbial hands in the mud: to count myself and see what that brings. That is, use or create a data set (or corpus) of mind maps, count the branches and leaves for each, and calculate the average and median. Since there are mind mapping tools, there must be files of mind maps that should be amenable to a computational analysis to obtain the answers, rather than having to go the cumbersome route of image processing and guessing, let alone manually counting the elements in each diagram and calculating the aggregates. I've done the dumb manual counting of models' contents before for another type of model and something useful came out of it,[5] which was a break in the feature importance

---

[5] They are conceptual data models, which we shall see in Chap. 4. The results of that counting are described in (Keet and Fillottrani 2015).

clouds, but it wasn't exactly an enjoyable task to do. To make a long story short for the mind maps: manually counting it had to be, of a dataset of mind maps that I had to create.

The prospect of manual counting was a constraining factor on the prospective size of the dataset. I settled on the 25 mind maps that the Learning Fundamentals website resources page showcases.[6] It may or may not be a representative set, but educational examples do nudge users in a certain direction of how things are supposed to be, and they are then less likely to deviate from it in a substantive way. Also, those mind maps cover a range of topics, from whole-food plant-based living, to the central nervous system, to getting ready for exams and to behaviour change programs, and thus a potential domain bias is evened out. Counting and averaging over their contents, then, the following emerged: each central concept has, on average, 6 main outgoing branches, ranging between 3 and 9 branches (median 6) and 29 leaves on average, ranging between 14 and 84 leaves (median 26). The maps with 3 or 4 main branches have the fewest number of leaves, but the few with 9 main branches do not necessarily have the most leaves. There are, on average, 5 leaves in a branch for each branch in maps in the data set, with the smallest average 2.5, in the mind map on getting motivated, and the largest is 9.3 leaves, which is the same map that has that whopping 84 leaves and is about 'creating effective change behaviour programs'. If your mind map is anywhere in these ranges of number of branches and leaves, you're in popular company. My mind map of dance has 6 main branches and 24 leaves, which comes close to the average. Yay. Also, these numbers are within the range of the sense of beauty for mind maps, according to Petr Kedaj and co-authors, with the Czech University of Life Sciences, Prague. Kedaj and co-authors also want a mind map design methodology, but instead of first devising a set of steps like those I listed above at the start of this section, they started their investigation at the other end, with what a map should look like. They need it to be balanced, rather than lopsided things with a starved branch here and a bloated one there, and without redundancies. Their notion of balanced is to have at most 7 main branches from the central concept, each branch should have a depth of at most 7 layers, and those leaves should be spread evenly.[7] There is no rationale for why they chose these values, however, nor an experimental evaluation. Notwithstanding, the teacher and computer scientist in me can see the advantage of such measures: they can be computed, and automatic feedback may be generated to dispense hints to the mind mapper to make the map look pleasing to the eye.

The mind maps in Buzan's book do look pleasing to the eye. Cursorily assessing them,[8] those maps mostly do have fairly balanced trees, and appear to have their average on the higher end of the range of those from the Learning Fundamentals resource. None of them go wild with 20 main branches and over 100 leaves, that's

---

[6] Accessed from https://learningfundamentals.com.au/resources/ and counted with the versions d.d. 15 April 2022.

[7] The metrics are described by Kedaj et al. (2014).

[8] That is, in (Buzan 2006), among the several books available by Tony Buzan on mind mapping.

for sure. If the mind map becomes that big, it's better to spawn off a new mind map for a big branch, or to modularise it manually. Moderating the size to the 6 or 7 branches at most lessens the chance of cognitive overload. And poring over a diagram that takes up the size of a dinner table is inconvenient anyhow. The world is not simple, however, and the further you go in education, management etc., the increase in complexity is unavoidable. To deal with all that, we'll have to shop for other types of models. We'll see those later in the book, as the difficulties are not insurmountable. Meanwhile, mind maps may assist in small brainstorming tasks and summarising chapters in schoolbooks, among its uses.

2.3 Limitations

While the activity of mind mapping has its uses and one can create pretty pictures with any of the apps, there are several shortcomings that cause it to be not nearly enough for a great many modelling tasks. We've seen the diversity in meaning of the branches: what comes at the next layer in the branch can be a disparate kind of thing, be it examples, properties of the thing at the parent branch, or parts or part-processes of it. Like in that example about a mind map on homework, with the examples of distractions and assorted components of the physical exercises. Yet, it's all shown the same in the diagram. There is a solution to that: diversify the notation of the branches. For some reason, this is not done unto mind maps. If you want that, you'll have to look beyond mind maps. Models in biology do just that, as we shall see in the next chapter (Chap. 3). If you were to think this might become complicated, it won't: you'll likely have been exposed to those type of models in primary and secondary school already, but now will see them with knowledgeable adult eyes.

It can get worse on ambiguity in mind maps. Take, for instance, some mind map about the school system, be it the one in your country, or a specific one where you or your children or your grandchildren or those under your care go. There would be a central concept **School** and a line to a leaf concept called **Learner**, where that line may or may not be adorned with a label. Let's be generous and assume the software allows you to name it: **enrolled-at** or **enrolls**, as you so prefer the reading direction to be. Could a learner be enrolled at two schools? Can a school have no learners enrolled? Must a learner be enrolled at least one school? Can a learner be enrolled at schools only or also other organisations?

We don't know. Well, actually we *do* know, but our map can't say anything about it. Does it matter that sort of information cannot be represented in a mind map? It does if you need a system to enforce any of those constraints or rules, because such constraints or rules would need to be documented for them to be enforced or at least to be communicated to the caregivers, whichever the constraints may be.

What could one do? One possibility is to adorn a mind map with a way to represent those constraints, be it by adding text or some icons. Another possibility is to relegate those tasks to a different category of models. We'll see one of those in Chap. 4, but first we'll introduce a set of models that allows more icons and has a clearer notion of modelling language and rules to govern it, in the next chapter.

References

Buzan T (2006) The ultimate book of mind maps. Harper Collins Publishers

Kedaj P, Pavlícek J, Hanzlík P (2014) Effective mind maps in e-learning. Acta Informatica Pragensia 3(3):239–250

Keet CM, Fillottrani PR (2015) An analysis and characterisation of publicly available conceptual models. In: Johannesson P, Lee ML, Liddle S, Opdahl AL, Pastor López O (eds) Proceedings of the 34th International Conference on Conceptual Modeling (ER'15). LNCS, vol 9381, pp 585–593. Springer, Berlin

Liu Y, Zhao G, Ma G, Bo Y (2014) The effect of mind mapping on teaching and learning: a meta-analysis. Standard J Educ Essay 2(1):17–31

Shi Y, Yang H, Dou Y, Zeng Y (2022) Effects of mind mapping-based instruction on student cognitive learning outcomes: a meta-analysis. Asia Pac Educ Rev. https://doi.org/10.1007/s12564-022-09746-9

Models and Diagrams in Biology

3

> ... the paradox that the more facts we learn the less we understand the process we study.
>
> — Yuri Lazebnik, in 'Can a biologist fix a radio?'

Mind maps on steroids. That's the minimum to call the group of models we'll look at in this chapter. They are loosely called 'biological models'. To be sure, it is feasible to use mind mapping to make biological models,[1] but then we're not taking a step forward in addressing some of its limitations. Biologists have better strategies. One way to make progress is to add additional types of elements and promote the drawing to serious icons. Biologists were heavy users of pictograms well before emojis became popular, however, and they like pictograms, too, like the mind mappers do. But more so, there's also the *freedom to choose* your elements! In several instances you can decide on what sort of pictograms to use in the diagrams. Compare that to mind mapping, where the drawing elements were just one text-filled oval at the centre plus the same thick lines with or without some text. If you want an arrow, say, you can add it now.

That freedom to choose and to add new varied elements in a flexible way in biology diagrams comes at a cost: it will take more effort to 'read' the diagram and understand everything in the figure. And there are a few rules to abide by still. Consider the pretty picture of the scientific understanding of COVID-19 at the end of 2021, as shown in Fig. 3.1. At first glance, there are a few superficial similarities with the mind maps: there are arrow-tipped lines between objects, as its own version of lines and pictures. So, for instance, instead of an oval with text saying "Dendritic cells" (at the top-left corner), we see a pretty sketch of how such a cell typically looks under the microscope. The lines do tend to have more meaning than their

---

[1] Mind maps on biology do exist, but only as part of educational material in primary or secondary education; see, e.g., Lu et al. (2013).

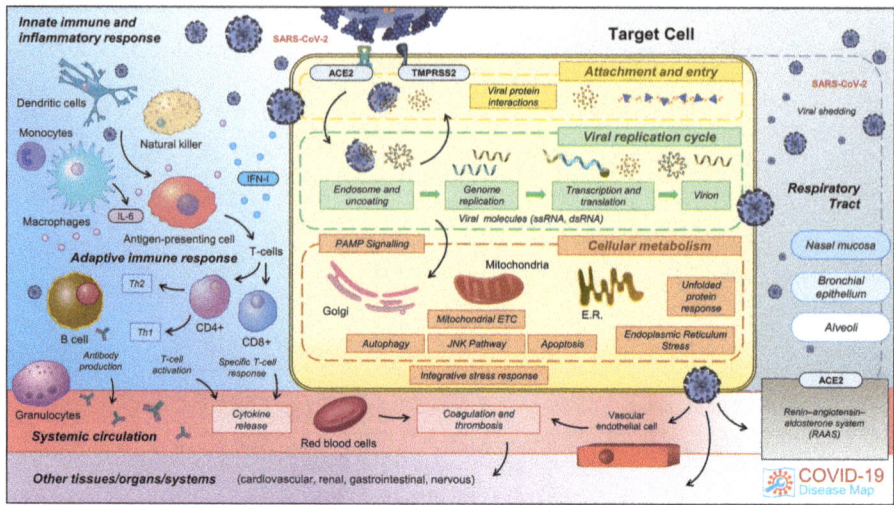

Fig. 3.1 Example of a rather busy diagram, about the COVID-19 disease. (CC-BY by Ostaszewski et al. 2021)

counterpart in mind maps. Don't worry if it doesn't make much, or only partial, sense at the moment—it will be straightforward by the end of this chapter. Also, the more courses in biology, chemistry, ecology and the like you've had, the more you'll have been exposed to how to observe its elements and how to read the diagrams. Many a reader may have dropped those subjects at secondary school as soon as they could and actively forgot about it. I noticed. I generally sneak a few biology examples into the courses I teach, and there's always a subset of students in class that visibly recoils in horror at the mere sound of 'something bio'. I don't know what the school teachers did unto them; anything can be made interesting. So, since this may apply, let's first look at how to read a biology diagram in a way that is, hopefully, not off-putting.

3.1 Reading a Diagram: Two Examples

Reading a diagram is the first step toward appreciating what's going on in those figures and, more importantly for our purposes, starting to discern regularities in them so that they become manageable regardless of the discipline within the life sciences. I'll introduce by example a number of key features of biological models from the viewpoint of principles of modelling. For those rusty on the biology side of knowledge, the examples will refresh your memory and will show it can be interesting. The first illustration is about alcohol fermentation—for drinks and for biofuel—and the second one is about who eats whom in the ocean among those creatures who go with the flow. We'll generalise from the key observations and bring in research on such diagrams afterwards.

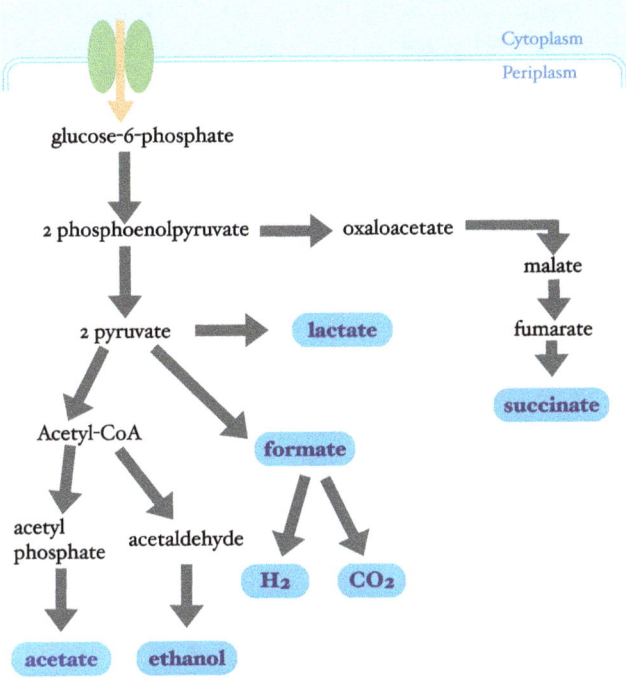

Fig. 3.2 Schematic diagram of mixed acid fermentation in *E. coli* (reproduced from the CC-BY image from Wikipedia).

3.1.1 Fermenting Sugars into Alcohol, Acids, and Gas

Let's start with an example on mixed-acid fermentation, as shown in Fig. 3.2. There's a Wikipedia page of 1417 words accompanying the figure to explain it all[2]—the picture says more than a thousand words. Or does it? It's about mixed acid fermentation, which is in the same family of processes as the fermentation of grapes into wine, hop into beer, and milk into cheese: a sugar is turned into alcohol, acids, and gas under conditions where there's little to no oxygen.

To start reading the diagram, you could start by making an inventory of types of elements, but that is not the most efficient approach. Just looking at the figure, what is the main thing that appears first? The key aspect that draws attention is those sequences of arrows. Second, there are those highlighted rectangles with names in them. Since the diagram is about fermentation, as mentioned in the indispensable caption of the figure, it has to do with chemical processes, so those names, such

---

[2] On 30 January 2022 the page had that many words, excluding references: https://en.wikipedia.org/wiki/Mixed_acid_fermentation, the figure's content being coherent with the "main compounds only" model on MetaCyc (https://biocyc.org/META/NEW-IMAGE?type=PATHWAY&object=FERMENTATION-PWY&detail-level=1 (last accessed on 29-5-2023)).

as the "glucose-6-phosphate" at the top, are molecules, as are those in the blue rectangles, such as ethanol—better known as alcohol and also used as biofuel—and H_2 and CO_2 as the gases. Those highlighted in blue are the important molecules in the fermentation: they're the ones that are produced eventually in this whole sequence.

Within that sequence of arrows, there are branches. For instance, "2-pyruvate" has three arrows pointing outward, which are typically directed downward or to the side and pointing to "acetyl-CoA", "formate", and "lactate" in the figure. This does *not* mean that it can take a random choice for whatever *E. coli* feels like producing today. Hence, something must be going on with those arrows, and indeed there is even though it's not mentioned in the figure. What's happening is a chemical reaction that is mediated by molecules called *enzymes* that make it easier for a chemical reaction to take place; an enzyme catalyses the reaction. The names of enzymes always end in *-ase*. In biochemistry or bio-organic chemistry, when there are diagrams like these, there are always enzymes that will get you from one node to the next, even if they are not mentioned in the diagram. For instance, that pyruvate can be used by the enzyme called lactate dehydrogenase and together with two helper molecules, NADH and H^+, converts it into lactate, which is also known as lactic acid. Lactate, in turn, can be used in other processes, such as in the production of bioplastics and as food additive for food preservation.

Likewise, the other two outgoing arrows from pyruvate use the enzymes pyruvate-formate lyase to produce formate and pyruvate dehydrogenase to produce acetyl-CoA. Put differently: the three outward arrows from pyruvate represent three distinct chemical reactions by three different enzymes.[3] It does not require a lot of stretching of the imagination to reason that it's likely not a random choice, but that it will depend on how much of those enzymes and helper molecules happen to be nearby to make those reactions take place. The choice can be nudged, too. Or pick another species of bacteria that will make those molecules in different proportions.

Anyway, this is about reading diagrams, not a crash course in biochemistry. So far, we first looked at the salient explicit things present in the diagram, and then added some implicit knowledge that an expert in the field will know about and therewith also will 'see' in the diagram. Tacit knowledge that experts have make them see more than a layperson, which is true for this example and for all those figures. For our current figure, what is remaining is the unobtrusive element at the top of the figure: those two green ovals with the orange arrow. Maybe it was the first thing that drew your attention, but it's not the exciting part of the figure. Those ovals over, or breaking up, a single or double line is an abstract representation of a protein that is in a 'wall', which is drawn with that double line, and it lets molecules through in one way or another, like a door in a wall in your home. This may be like just a door opening as if it were an automatic door or it may only open for payment with the right key.

---

[3] MetaCyc allows a user to select five levels of detail, at: https://biocyc.org/META/NEW-IMAGE?type=PATHWAY&object=FERMENTATION-PWY (last accessed on 29-5-2023).

3.1 Reading a Diagram: Two Examples

Regarding the wall, since walls exist in several places, there must be an indication which wall the figure is referring to. In this diagram, there's "Cytoplasm" and "Periplasm" written, with the chemical reactions taking place on the periplasm side. But where is it? You may have heard of cytoplasm, which is the substance that's inside a cell. Periplasm is a much less common word, but rudimentary Greek and Latin, analogies to familiar words, and basic principles of space come in handy in deciphering and guesstimating what it means. Since *E. coli* bacteria are single cell organisms, the chemical reaction must take place in, on, around, or just outside the cell. The cyto- in the cytoplasm means 'container' or the 'receptacle' and while there quite a few words in the English language that begin with cyto-, none of them are often used. That's a different story for exo-something, such as exodus, exoplanet, and exotic; 'exo' means outside or external. Since peri- is neither cyto- nor exo-, there's only one option left really: round, roundabout, near, like in periodontist (around the tooth), perimeter, and peripheral.[4] Periplasm is that plasma that's around the inner part of the cell that contains the inner plasm. That only works if there are two layers of cell walls. And indeed, bacteria may have that, but only the so-called Gram-negative bacteria have an inner and outer membrane layer. This contrasts with Gram-positive bacteria, which lack that outer membrane and therefore they stain purple instead of pink in a test called Gram staining. Mr Gram developed that test. So now we know not only where these processes happen—the periplasm—but also in which group of bacteria. All that gleaned from one 'simple' figure.

The more you learn about some domain, the more you see in a diagram. Learning to read a diagram means learning about the subject domain, and vice versa. It's both a strength and a weakness of these biological models. A strength for communication among experts. A weakness due to all the implicit knowledge that's not there but essential to understand what's going on in a diagram. It's a bit of a catch-22. It's cumbersome to write about reading diagrams and, to the best of my knowledge, it's supposed to be imparted by a teacher in school.

3.1.2 Who Eats Whom in the Ocean, at the Microscopic Level

The biochemistry diagram is structured and not one of the 'cartoon' type of diagrams, so let's look at one of those and surprise ourselves with the implicit systematics of it. Instead of the sub-cellular level, at the other end of the spectrum in ecology, it's not any easier because the things are larger and observable with the naked eye. We'll see it's about the same when it comes to the interaction between the subject and learning to read the diagrams. A simplistic-looking representation of a microbial loop in the ocean is shown in Fig. 3.3. The arrows here mean something else compared to the arrows in the figure about the alcohol fermentation. There are

---

[4] But not the peri- in peri-peri, a popular type of sauce in South Africa: etymologically, it's traced back to the Swahili *piri-piri* where *piri* means 'pepper' and with reduplication indicating emphasis or superlative, peri-peri means 'very peppery', or spicy.

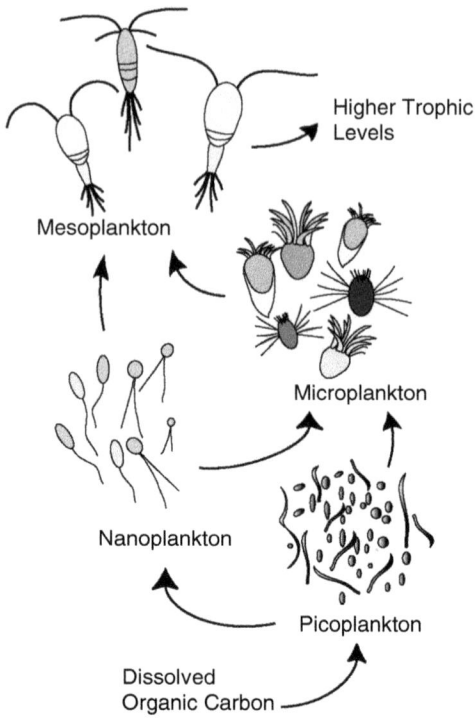

Fig. 3.3 A simplified microbial loop (public domain image from Wikipedia)

no enzymes that catalyse chemical reactions: the organisms depicted at the tip of the arrow *feed on* whatever is mentioned at the start of the arrow. On those names: they're all "plankton" things, which is a loanword from German that, in turn, had borrowed and adapted it from the Greek *planktos*, which means 'wandering' or 'drifting'. Now it refers to a group of species that float wherever the current in the sea takes them; that is: they're organisms that can't swim anywhere they might want to. They include, among others, bacteria, green algae, krill, and sea snails. For the pico, nano micro, meso, and zoo in the names: it's mostly Greek again, where the first four indicate size. Picoplankton are drifting organisms between 0.2 and 2.0 micrometer (μm) in size, nanoplankton's cells are between 2 and 20 μm, whereas microplankton is anywhere between 20 and 200 μm long, or at most 0.2 mm, and mesoplankton are organisms or clusters of them between 0.2 and 2 mm in size. Picoplankton grazes on dissolved organic carbon, microplankton eats picoplankton, mesoplankton feeds on microplankton, and 'higher trophic levels' feed on mesoplankton. That 'higher trophic levels' starts with zooplankton that, in turn, are eaten by other, larger, zooplankton or larger animals. This happens in water—deemed to have been obvious enough to omit from the diagram.

Something else is going on in the figure, which is indicated in two ways, once you know what to look for. First, the pico-, nano-, micro-, and meso-, is going from the very small to the less tiny, which is indicated both with the names given and it's shown graphically with bigger objects as we go along upward in the food chain. The plankton are not drawn to scale, but the idea is there. Second, each xxxoplankton refers to organisms of diverse species that are within the specified size range. Now look again at Fig. 3.3: the 'diverse species' aspect *is* indicated, with the distinct figurines at each end of the arrow. The person who made the figure wasn't just randomly doodling one or two tails for nanoplankton and circles, ovals, and worm-like shapes for picoplankton as a just-for-variety to make it all look more interesting; they denote organisms of different species. Even in cold seas that freeze over winter, such as the Baltic Sea, there are well over a thousand "operational taxonomic units"—probably species, based on the sequenced ribosomal RNA—of bacteria and close to 2000 operational taxonomic units of eukaryotes (plants and animals) drifting around in the water.[5]

Last, cognitively and in speech, one goes 'up' in the food chain, and so there is that vertical representation. The 'loop' aspect is not drawn. It could be, but then there would have to be an arrow downward from the "higher trophic levels" to the "dissolved organic carbon": the organisms in the sea die, decompose, and, voilà, there is our organic carbon.

Thus, also in this second example of a 'cartoon-like' biological diagram, it relies on substantial background knowledge about plankton to be able to observe the key aspects of the figure. Describing it ends up in a two-way interaction also in this case: the more components you see, the more there is to grasp about what is really going on, and, vice versa, the more knowledge you have about the subject domain, the more there is to recognise in the figure and to appreciate its subtleties. The scientific models of the microbial loop contain further information, refining both the organisms partaking in the loop and the molecules passed along.[6]

Are all the models about processes that things participate in? No, but very many indeed are. There are models with just structural things, like the components of a cell, the parts of a plant, or human anatomy. A model of the latter, the so-called canonical human body, has well over 200,000 kinds of things recorded, from the assorted bones in your body to the organs, types of tissue, and so on, but there just are many other processes happening in and with a human body. For the humble cell, the most widely used reference source, called the Gene Ontology, has about 4000 named cellular components and a whopping 28,428 cellular biological processes—and counting and documenting them still since 1998 when they started with the endeavour.[7]

---

[5] See (Hu Yue et al. 2016).

[6] Refinement of the organisms in the loop includes protozoa, bacteria, phytoplankton, zooplankton, and algae; the molecules traced includes notably also nitrogen in various forms. One of those models is described by Tett and Wilson (2000), which was assessed in the context of the topic of Chap. 5 and is described in (Keet 2005).

[7] The Foundational Model of Anatomy is the most well-known reference source that not only lists all those components, but also relates them (Rosse and Mejino Jr 2003). The Gene Ontology statistics apply to the January 2022 release, available at http://geneontology.org/stats.html; the ambitious endeavour was first described in 2000 in (Gene Ontology Consortium 2000).

It merely may just feel worse: the snag with the components is that the relevant subset has to be memorised. They hardly change compared to the research about all that is happening and therefore lots about that is implied in any biological diagram. That cytoplasm and that periplasm in the diagram? It's been the same for very, very many years. As would the mixed acid fermentation be, mostly, true, but those processes can vary depending on species and context and manipulation of the organism's genes, which is why such processes receive prominence in the diagrams. Likewise, new diseases come along that have new processes with existing structures, existing processes are gradually better understood, and processes change due to climate change, and so on. For one new virus, SARS-CoV-2, there's that fascinating disease map in Fig. 3.1 with all those processes in that orchestration specific to COVID-19. Counting the number of arrows, as an approximation for processes happening, is left as an exercise to the reader.

3.2   A Quest for Common Characteristics

There are so many of those diagrams in the sciences that it made people wonder about that state of affairs. Why not just read text? Is there something that can be shown in a diagram that cannot be expressed in natural language? Are they easier to understand than text? Do learners learn the material faster with those diagrams? What are the commonalities across textbook diagrams? Can a classification of diagrams be made? Can one reason over the diagrams? Is there some fundamental difference between manually drawn diagrams and those drawn with diagramming software that limit drawing flexibility? Are there rules for drawing diagrams? Questions, questions, lots of questions. There are answers.

As we have seen in the fermentation and plankton examples, arrows are directed from the 'predecessor(s)' to the 'successor(s)', if there's an entity at both ends and those things are distinct things. Predecessors can be things like molecules, cells, or organisms, as will the successors be, and often be in the same category of things as its predecessor. If there's nothing at the other end, either the thing on the predecessor end moves to where the arrow points, or the successor came from the place where the arrow originates from.

One would think it should be attainable to automatically analyse diagrams on what is in them and what the most common elements and patterns are, considering there are all those algorithms to detect cats on photos on the Web and optical character recognition after scanning pages of text. We're not there yet. It's tempting to speculate that it may be because engineers prefer cats over biology. We'll look at more probable explanations why there are no good algorithms for it yet in the next section, as well as attempts at rules for drawings that assist in determining whether a diagram is good or bad and why.

3.2.1 The Chemists and the Cladists

Certain disciplines and even sub-fields within a discipline may have specific rules for a diagram to be a valid one. Their practitioners got together and formed societies and created diagramming and standardisation committees. A discipline that is very organised and strict with willing adherents, is chemistry. There is, firstly, the International Union of Pure and Applied Chemistry (IUPAC) since 1919, which defines rules for nomenclature and symbols. Biochemistry branched off in the late 1940s and set up an organisation in 1955 to increase its influence, called the International Union of Biochemistry and Molecular Biology (IUBMB). They focus on enzymes and metabolic pathways, like those we've come across in the mixed acid fermentation. Remember that lactate dehydrogenase? It's identified with EC 1.1.1.27, with the "EC" referring to the Enzyme Commission of the IUBMB and each number from left-to-right is a more specific group of enzymes. One might think that pharmacology, which clearly also has something to do with molecules and biochemistry, is within the realm of either as well, but no, there's also a union for that, the International Union of Pharmacology (IUPHAR). They started in 1959 as a section of yet another union, the International Union of Physiological Sciences, and then went their own way as IUPHAR in 1965. They're in charge of the Receptor Nomenclature and Drug Classification.[8] But coordination in naming is not yet rules for drawing.

Fortunately, it turns out that it's roughly the same story with diagrammatic notations for chemicals, which are called a *structural formula*. By now, every textbook uses the same notational conventions in the figures. It didn't use to be the case. There were various attempts at drawing them in the decades that molecules became fashionable to be investigated. Over a century ago, Gilbert Lewis, then a physical chemist with the University of California in Berkeley, USA, discovered the covalent bond between atoms and the notion of electron pairs and he needed a way to draw that in his scientific articles. The way he did it in that key publication of 1916 has become known as the *Lewis structure* notation. It has been extended and refined over time, but the core notions are still recognisable and the name stuck.[9] The most adorned ones adhere to the *vsepr* notation, which is an abbreviation for 'valence shell electron pair repulsion'. In plain English, it means that vsepr tries to make the abstract representation of the molecule look a little bit 3D-like. The bond between atoms is depicted as a line, with the atoms denoted with their approved abbreviation (H for hydrogen etc.), double bonds with two lines, and so on. The 3D-like feel is indicated with solid wedges, indicating that the atom at the thick end is to the fore of the one at the point, and dashed wedges indicate that the atom at

---

[8] They can be accessed online at, respectively: https://iupac.org/, https://iubmb.org/, and https://iuphar.org/, respectively. (last accessed on 29-5-2023).

[9] The article by Lewis (1916) is limited compared to what now passes for 'Lewis structure', notably missing the lines, and a single dot is only used when there is a free radical, with double dots replaced by lines between the atoms.

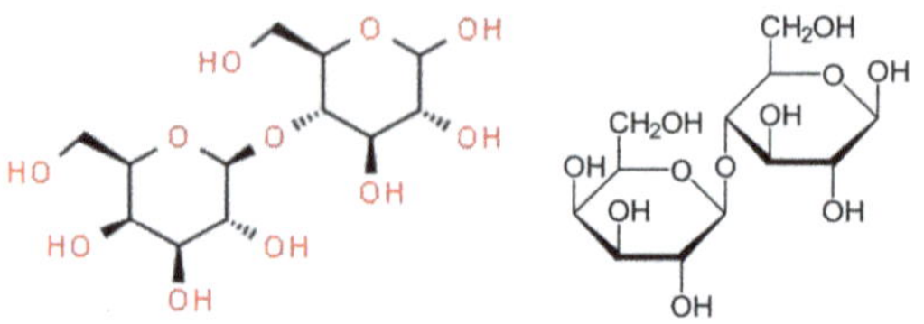

Fig. 3.4 Lactose, the sugar to which many people are intolerant: their digestive system doesn't produce sufficient β-galactosidase enzyme and so the gut bacteria can feast on fermenting it instead. Diagrammatically in two permissible ways: Left: the Lewis notation, where those wedges suggest the element attached is either to the back or to the front of the flat hexagon; Right: Haworth notation

the tip is positioned to the back. An example is shown in Fig. 3.4. There are further conventions for specific groups of molecules, such as the so-called chair notation and Haworth projections for sugars.[10] There is only a small set of such approved graphical notations for molecules and all molecules can be drawn with it.

Diagrams of chemical reactions, especially relevant in biochemistry and bio-organic chemistry, all adhere to those rules for graphical notation of the molecules. The reactions themselves are depicted from left to right or top to bottom, or both. There may be an equilibrium, where the reaction also happens in reverse, so strictly it then also goes to the left, but that is not a common situation. The flexibility on how the knowledge may be represented in the diagrams is limited. This put the door ajar for a whole host of attendant activities, such as automatically analysing such diagrams, devising methods for their development, clarifying notations also in writing as opposed to relying on the oral tradition (teaching in class) for passing on reading the diagram, and so on.

It isn't necessarily less organised everywhere else. The cladists—those who categorise organisms into 'groups', also called clade, taxon, or species—like their order as well. It's their life's work to systematise organisms. Their discipline of cladistics, or, accurately, phylogenetic systematics, has an organisation to promote it, called the Willi Hennig Society.[11] Unsurprisingly, they also have come to an agreement about how exactly those diagrams must look. There are evolutionary diagrams, and then there are *cladograms*. The cladists make cladograms. Now is also the time where the notion of good and bad diagrams enters the scene, which we didn't need for the diagrams in chemistry. The chemists agree for many decades and are top-down oriented, which has resulted in a state where there aren't any bad chemistry diagrams that violate notational conventions. They're all good when it comes to notation. The cladists, however, are still fighting the good fight to get everything and everyone in line, which is harder to do with the Web being pervasive

---

[10] An online free resource that has a longer summary explanation of the notational conventions of structural formulae can be found in (Riehl 2010).

[11] Information about the society is available on their website at https://cladistics.org/ (last accessed on 29-5-2023).

and an earlier situation without clear rules. Now there are such rules. Not only rules for rules' sake, but they have a theoretical basis reflecting scientific insights. Violating a rule, then, entails missing a point of the science of it. Not even all schoolbooks get it right.

With the rules in hand, Kefyn Catley of Western Carolina University and Laura Novick of Vanderbilt University in the USA investigated evolutionary diagrams in biology textbooks. In describing their analysis—and lamentations on the sad state of affairs—they also impart the rules for such diagrams and, importantly, why the other ones are wrong and based on misconceptions.[12] What's going on? How can it be that bad? Let's start with the 'tree of life' that you undoubtedly will have seen somewhere and certainly heard of. It has bacteria at the stem or bottom of the tree, and then lines fanning out in multiple directions, with species names for groups of organisms, and perhaps even cartoons of species along those lines, and again names at the end of those lines. They are typically pretty pictures, especially when they're in colour. They're also wrong, according to the cladists. Those misguided diagrams communicate, at best, outdated ideas and don't reflect current scientific understanding of how the species relate.

Enter the reasons *why* proper cladograms are drawn in one way and not another. In a cladogram type of diagram, the lines schematically represent evolutionary time albeit not necessarily at scale, the nodes at which the lines fork indicate a common ancestor, and at the end of those lines there's a species name for the group of organisms and those ends are all at the same height. The lines may be adorned with an annotation what the feature was that had it split, such as having two claws instead of three, or feathers versus fur, say. Anything else is not good. A line where there are a few species along the way? Bad. Speciation is a branching event, so one line must split into two in the diagram, rather than the anagenesis where one species would have kept interbreeding and gradually turned into another along one single line. Naming the nodes in the tree is less bad, but because it's still somewhat suggestive of anagenesis, it shouldn't happen either. Those trees, influenced by German zoologist and evolutionist Ernst Haeckel, implicitly communicate teleology, i.e., as if it were to be purposeful evolution from the 'lower' bacteria toward the 'higher' animals, which is also not how evolution is understood—adaptation to the environment it is. What about drawing a side branch so that there are more than two outgoing lines? Nope. They violate the notion of monophyly: only two taxa (species) share a most recent common ancestor. If there's a third species, it must have branched off earlier or later, but not at the same time. The lines also must have the same width, since it's not like there were always a larger number before and fewer of them now, nor vice versa. I've illustrated a few aspects in Fig. 3.5.

There is still a certain level of freedom for how to draw them, like a so-called ladder, which is a tree with diagonal lines, or one with lines at a 90° angle vertically or horizontally, and from bottom to top or left to right, or even in a circle-shape

---

[12] The research described in this and the next two paragraphs, together with a discussion of wrong diagrams and why they are wrong, is based on the work by Catley and Novick (2008).

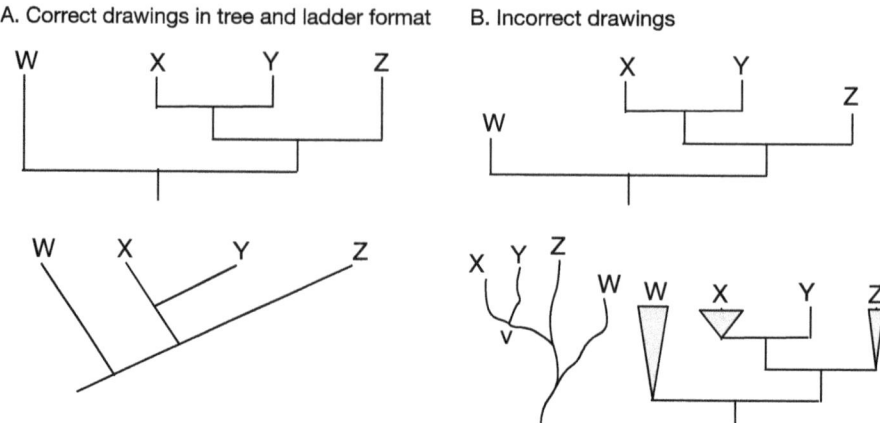

Fig. 3.5 Stylistic illustrations of the structure of good cladograms (left) and incorrect diagrams (right)

(see Fig. 3.5a). Colour may be added, and the leaf nodes may be adorned with pictograms, too. The systematic biologists overwhelmingly prefer trees over ladders in their scientific papers. Textbook authors more often go for ladders—and other diagrams that violate the cladogram rules.

Returning to Catley and Novick's investigation with the rules now in hand, out of the 192 diagrams across 31 textbooks, they found a whopping 28% of the diagrams having one or more errors that indicated misconceptions of the science, like those described in the previous paragraph. Maybe textbooks are slow in the uptake of the latest research, one might suspect. So I cherry-picked cats as clade and assumed a persona of learner without textbook or someone who asks a search engine about the evolutionary tree of cats. It steered me towards Wikipedia and the first openly accessible popular science article. The enjoyable read about the evolution of cats in the Scientific American popular science magazine has a tree like the top one in Fig. 3.5b, on the incorrect side of the matter. Wikipedia's entry on cats, the *Felidae* family, does not fare better, but reveals an interesting situation: the source it cites has it right and the blame lies with software choices of Wikipedians. That tree representation about cats where the species names are not aligned is due to the use of a template that has been instantiated on about 7500 Wikipedia pages, including the cat page. The template repurposes HTML table specification features to approximate the so-called Newick tree format, rather than cladograms that are a different type of tree. Considering how HTML tables on webpages are specified, it's certainly easier this way compared to aligning the names to the right.[13] Yet, one ought to be mindful about what one is drawing.

---

[13] The Scientific American article was written by Driscoll et al. (2015). The Wikipedia page on cats at https://en.wikipedia.org/wiki/Felidae references Li et al. (2016) for the large cladogram. The template is available at https://en.wikipedia.org/wiki/Template:Clade (last accessed on 29-5-2023).

3.2.2 One Diagramming Language for all Biological Models

What if there would be only one set of icons to choose from and only a limited number of ways to put them together? Then it's effectively a language, like English that has a vocabulary and a set of grammar rules. At first thought, this may sound appealing; however, the fact that we needed a few pages of text just to describe the elements in two diagrams of the examples about mixed-acid fermentation and plankton does not bode well for such a universal diagramming language, and then we might as well just write it out in plain English itself. A key benefit of the diagram is that it's intended to capture more than a thousand words, literally.

There are potentially two ways out of this conundrum: either make a diagram language for one subject domain only or figure out what the most used icons are and force everyone to use those, possibly with refinements. The first strategy has as two main upsides the fixed set of diagram elements and grammar rules for their use, but the downside is that for each new sub-field, you'll have to learn a notation and even within a sub-field there may be multiple competing dialects to deal with. This strategy is amenable to a computational approach: computing has a sub-field called Domain Specific Languages that has lots of software infrastructure to support exactly that. The domain experts decide on the key ingredients that are to become the core types of elements that can appear in a diagram with an icon for each element (the primitives), and they specify how those elements can interact in a set of rules (the grammar), and voilà, there will be your language. For cladograms, such a Domain Specific Language would have:

- Elements: line of flexible length, binary fork (one line in, two out), text box, short thick line, and image;
- Set of composition rules to ensure it results in a correct cladogram:
  - line ends of a binary fork must be aligned;
  - each line end must have either a non-empty text box or an image;
  - an image may appear only at a line end;
  - binary forks may be concatenated at either of the two 'out' ends, except when there's a text box or image at the line end; and
  - a perpendicular short thick line may be added only at the branching end of a fork and needs to have a non-empty text box associated to it.

Figure 3.5a satisfies them all, though noting they don't have those perpendicular short thick lines. It's quite doable for a computer science graduate to write a program to render such element and check they adhere to the constraints. An added benefit would be that verification of correctness of the layout of the diagram then also can be left to the software that can return feedback to the user on whether the diagram adheres to the specifications or not. For instance, if the short thick line doesn't have text yet, the tool could draw the user's attention to it with a red question mark and display feedback that text still has to be added, or if the user drags an image to a fork, it will be repelled (not allowed to be dropped there) because a fork is not allowed to have an image. It can work analogously for the fermentation processes, or for the microbial loop, and any sub-group of models in a field of specialisation. And as many apps to develop, install, and become familiar with.

This brings us to the second option, that small subset of most common icons. It may emanate from developing all those domain specific languages and tools to generalise from that, or from an attempt to specify it upfront and disseminate to the domain specific languages. The upside of this option is that once you know those most used icons, you'll be able to interpret any diagram. The downside is that there's no such list and it is not clear what deserves to be on it and who decides about that. Except for arrows. They're definitely essential.

This theoretical foray may start to feel like a 'and what's out there that does all this for us already?'. I did conduct a literature search on it several times throughout writing the book. Surely there ought to, nay must, be at the very least a proposal already on those generic elements?! I know of a few domain-specific tools since the mid 1990s when I was exposed to it when writing up my thesis in microbiology. The one I was most impressed with was ChemDraw, which hails all the way back to 1985, the stone age of graphical user interfaces for end users. It was the first ever chemical drawing software application and one that built in certain constraints and the standard notation.[14] But alas. There are a range of tools, but research and science-based evidence on diagramming and good user interfaces for it is sparse, if existing at all.[15] A scientific angle to diagrams might sound far-fetched but is not. Consider the notions of readability and draw-ability of diagrams: certain element shapes and colours are more easily associated with one thing or another simply by societal priming. For instance, one may associate the representation of a protein readily with an oval rather than a rectangle, and up-regulation with green (indicating growth and going ahead) rather than orange (of a traffic light strongly suggesting to stop and wait). Scientist do research on that sort of thing, blending computing, IT, psychology, and cognitive science in a field of specialisation called Human-Computer Interaction, but not so much when it comes to biological diagrams.[16]

Curiously, or oddly rather, a lot of research is being carried out on wholly disparate questions for diagrams in biology. They essentially amount to 'do learners

---

[14] Halford (2014) chronicles the tool from its inception by Stewart Rubenstein and Sally Evans to its 30th anniversary in 2014. The current website URL of the tool is https://perkinelmerinformatics.com/products/research/chemdraw (last accessed on 29-5-2023).

[15] The following examples are just that, and not an endorsement. There are tools for all sorts of modelling, of which science diagrams are one group of models in their suite, such as those by Edrawsoft and Edrawmax, and repositories of diagrams, such as SciDraw; their respective websites are: https://www.edrawmax.com/science-diagram/science-diagram-maker/, https://www.edrawsoft.com/science-diagram-software/, and https://scidraw.io/ (last accessed on 29-5-2023).

[16] A search on 'biology' in the HCI journal from Taylor & Francis yielded 59 hits of which neither was relevant; there were more hits in the International Journal of Human-Computer Studies (on 17 July 2022), but mainly also irrelevant with respect to the topic of alternative diagrammatic representations. It is not unheard of: an example is our experimental approach to temporal notations of ER diagrams that we'll see in the next chapter (Keet and Berman 2017).

understand these diagrams?' and related attempts at sense-making by non-biologists trying to figure out about what exactly those biologists possibly could be saying. As if aliens dropped the diagrams on earth or they sprouted from the cabbage and those unknown entities need to be prodded and poked and turned around and dissected. The diagrams with their notations are human creations, however, so one could ask the people who create them instead of tiptoeing in a roundabout way. Not just that, one could structure the visual languages, develop methods to create good diagrams, and tell the readers what the icons and the rules are. That would make the sense-making for educational material a whole lot easier.

In sum, while it may look from the outside as if it's a Wild West in diagrammatic notations, there are some commonalities among them and efforts at structuring the notations, even if mostly implicit or as tacit knowledge one is expected to acquire during one's education. The diagrams often contain a certain process view of the topic, with any static aspects in the background and then only limited to the minimum information needed to determine the location where it is happening. The arrows indicate time, in sequence or in parallel if they're branching, where things either move there or happen, like a reaction, eating, or flowing to another place.

3.3   How to Create a Biological Diagram

The section's title is more of a question than a statement since there are no settled methodologies for how to create a biological diagram. Like with mind maps, it's quite workable to draw biological diagrams with pen and paper, like the microbial loop in Fig. 3.3. There are also numerous general-purpose and special-purpose applications, and there are lots of stock icons online that can be reused both from public domain resources as well as those that come with the diagram drawing software. But opening that application or taking that sheet, you have a blank canvas. Where do you start drawing? Or, more specifically: *how* do you start drawing? Is there a procedure for doing this? A procedure not in the sense of a user manual for a tool or advertisement thereof, but *irrespective of the tool*.[17] That is, some sort of cookbook-like instructions that will work throughout the years irrespective of the technology. For instance: should you start drawing the 'context' first? The objects that undergo change? The sequences first and then find the pretty pictures to swap them with? Could the order of actions possibly matter, and if so, in what way? During my times in microbiology and cell biology in the 1990s, there weren't any methods for how to go about doing this. I can't recall I was looking for a methodology to do that, either. I just went ahead and started drawing. It's a long

---

[17] To prove the point, search for "how to create a biological model" in the most popular search engine. I did, twice, which returned as first hit an advertisement plug in the prestigious journal *Nature*, for the Virtual Cell software at https://vcell.org/index.html (Kritikou 2007), which, perhaps unsurprisingly after some 15 years, suffered from link rot (i.e., a 404); https://vcell.org/ works (last accessed on 29-5-2023).

time ago. The only reason why it occurs to me now to question and look for methodologies, methods, techniques, and tools, is because we have them for the conceptual models we'll see in the next chapter and they have been shown to be useful in the sense of improving quality, precision, fewer errors, or modelling faster. Conversely, an absence of methodologies is a sure way of forgetting to add things, adding the wrong things or in the wrong way because you don't have the rules at hand, and thinking longer about the elements and the rules because they're not systematically followed.

Sadly, in the timespan of a whole generation, there still don't seem to be any methods, and it's thus also not known whether a methodical approach to modelling would result in better diagrams. I searched the literature, but to no avail. Relevant-sounding books focus on quantitative aspects rather than qualitative modelling.[18] as if the latter comes naturally. Yet, it would make sense to lay down the qualitative aspects—what is involved in what way—systematically before adding the mathematical formulae and devise simulations, even if only for documentation and comparisons of models on their constituents and underlying assumptions rather than only comparing outcomes. At the time in the late 1990s and early 2000s, such modelling, if done at all, was an activity you do on the side, emanating from *the real research* done in the lab. This has changed in the meantime, in that in silico, or 'on the computer', research has become respectable.

This lack of guidance on how to do the modelling is in stark contrast to the modelling in computing and IT. Since I have acquired that knapsack of knowledge in the meantime, I might as well cherry-pick from it to design a procedure for developing biological models. A first attempt would be as follows, for illustration:

1. Drop the images for the nodes on the drawing canvas for those that have distinguishable images/icons and write the name of the others;
2. Draw the arrows between the elements to create the principal layout;
3. Add names to the lines and nodes (images), where applicable;
4. Find any remaining images/icons, if needed and add those;
5. Draw the context around it, with minimal hints of where it takes place, where applicable;
6. Do any further colouring in and layouting to make it visually appealing and easier to read;
7. Add a caption such that the whole (caption + diagram) is self-contained and explains what it is about.

Purely scientifically, I must emphasise that this sequence counts only as a suggested *procedure*, not a *methodology*. For it to possibly become a methodology, it has to be scientifically evaluated first.[19] Be that as it may, let's take those cladograms as an example for tailoring this procedure to a specific type of diagram. You could

---

[18] For instance, the very recently published Biological Modeling, at https://biologicalmodeling.org/ (last accessed on 29-5-2023).

start at the base, i.e., the common ancestor to all the taxons to be included in the cladogram and work your way outward to the taxons at the leaves. Alternatively, the other sequence starts at the leaves with the taxons to be included and then you work your way back to the most recent common ancestor of all of them. Here's a sample procedure for the second option:

1. Write all taxon names at the top of the canvas, aligned at the same height;
2. Optionally add images for those taxons;
3. Pick two evolutionary most adjacent taxons, draw a line downward from each taxon to make them meet to create a new node;
4. Repeat the previous step for the taxons and the nodes until there are no more loose taxons or nodes;
5. Draw a little line from that bottom-most node further down;
6. Optionally, working your way upward from the base, add an annotation on the line before and after a node with the trait (called the synapomorphic character) that the two taxons had in common and made it split off, respectively;
7. Finalise the layout such that the tree looks like an unbalanced binary tree or a ladder;
8. Add a meaningful caption.

Having a procedure like this avoids mistakes that Catley and Novick had observed. For instance, there is no step that says 'add name to an intermediate node', and so they will not be labelled, and there is a step that says 'align at the top', ensuring that the taxons at the leaves are not all over the place. A methodology for diagram design can avoid bad diagrams. Anyone with an inquisitive mind who reads this procedure can formulate questions easily, such as 'why align at the top?' or 'can I add names to the intermediate nodes as well?', which then serves as easy entry points to clarify concepts of the science summarised in cladograms. Perhaps such a methodical approach to modelling is uncommon, but research in biology is rife with protocols for wetlab and field research already. A few more protocols could go on that pile.

What about our running example with dance? Not as easy as a mind map would be an understatement. Also, the mind map in Fig. 2.3 has no content on anything biology. The closest that comes to something about dance itself, is at the top-left corner of the figure: partnerwork and larger configurations. There's a myriad of notation systems for steps and body movement,[20] but there is no biology to

---

[19] The key questions to answer in such evaluations are whether the proposed procedure has better results than the status quo (*in casu*, no official methodology). This entails devising a measure and a way how to measure what 'better' means, and an experimental evaluation with a large enough target group of people to compute statistical significance that the models made with the procedure are better.

[20] They are amenable to computation and two of my honours (4th-year) students in computer science tried that. Their code and reports are available from the honours projects archive, at https://projects.cs.uct.ac.za/honsproj/cgi-bin/view/2019/baijnath_chetty_marajh.zip/

it. Movement and coordination and synchronised dancing is good for the body, but then the real models are those of the physiology and neurological processes of the human body. There is also the evolution of dance across species, superb lyrebirds' coordination of their song and dance, how bees do it and much more. To illustrate the diagram design procedure and that it may improve a diagram, we shall take the superb lyrebird's song and dance, of which it's claimed to be the first evidence that non-human animals have a dance repertoire to go with distinct types of songs, and they manage to do both at the same time. Anastasia Dalziell from the Australian National University in Canberra and colleagues took several trips to Sherbrooke Forest near Melbourne to take videos and audio recordings of singing and/or dancing superb lyrebirds (*Menura novaehollandiae*), 12 of which with full dance display performances. The recordings were analysed for the audio frequencies and durations to detect songs and the videos for movement. A combination of song and movement was deemed dance if the movements weren't utilitarian like walking and flying, but to show off, like doing steps, jumps or bobs, or a wing flap that wasn't about flying. It turns out they perform quite a show.[21] It is that show—a sequence of songs and dances—that the authors modelled in a diagram, which is included here as Fig. 3.6. The quality of the diagram easily can be improved upon with the aforementioned procedure. First, the attention to images and icons. The boxes labelled 'Start', 'End', and A-D neither indicate difference nor what they represent, i.e., not exploiting step 1 of the procedure, nor does the caption clarify that beyond size indications. It's easy to stylise the article's Figure 2 on sound and movement to create visual distinctions between the four song and dance pairs, and to distinguish it from start and end, so we already arrive at Fig. 3.7. The video abstract of the article also has a nice visual combination of the song and the dance at minute 4:07, although reusing that would entail abandoning the box size difference as visual clue. The arrows don't have names, but there are probabilities labels in their stead that may suffice. A distinctive colour for each of the A-D boxes could further emphasise the differences.

Comparing this with the mind map about dance from the previous chapter, the first observation is that the elements carry more information, and the information is more systematically of the same kind of thing. It's not that a line could mean anything, but it denotes a transition from one action to another and with a probability associated to it. Thickness of lines and the size of the boxes entail relative importance rather than mere aesthetic. The oval and box shape each is of a category of meaning, rather than labelling a same type of branch element where anything can go. Whether all this is *better* is a separate question that we'll get to answering later.

DEDANCE_website/ and https://projects.cs.uct.ac.za/honsproj/cgi-bin/view/2021/dauane_kriel.zip/dauane_kriel/ (last accessed on 29-5-2023).

[21] The discovery by Dalziell et al. (2013) is described in a freely accessible article and has a great video abstract where the birds can be seen in action, at https://www.cell.com/current-biology/fulltext/S0960-9822(13)00581-2 (last accessed on 29-5-2023).

3.4 Limitations

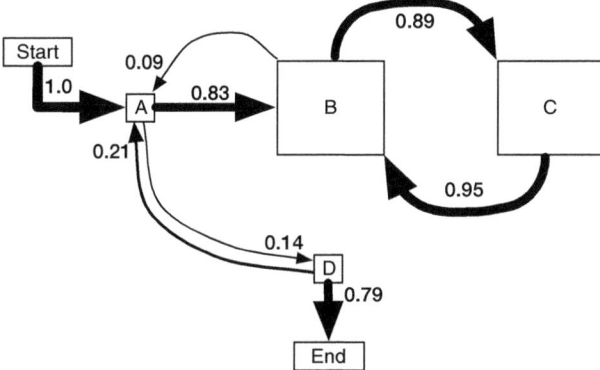

Relative Frequency of Occurrence and Sequential Relationship of Song Types during Dance Displays

Displays start with song type A and often end with song type D. Transition probabilities suggest the following typical sequence of song: from song type A, they alternate between B and C several times before repeating song type A and finishing with song type D. Song types B and C were performed at least four times more often than A and D song types, as indicated by box size. [...]. Transition probabilities less than 0.05 are not shown.

Fig. 3.6 Recreated diagram of the superb lyrebird's show, with a part of the original caption added (Reprinted from Dalziell et al. 2013, with permission from Elsevier)

3.4 Limitations

Let's first consider the good news. Biological models can address the shortcomings mentioned for mind maps. There is much less diversity in meaning of icons in the diagram thanks to being allowed to diversify the icons and notation of the lines into the various arrows. Lines many be annotated with names and percentages and the like. There is even a regularity to notations across models in different disciplines within biology and there do exist roughly standardised notations for some. We have gained much expressiveness.

Yet. While there are rules and pre-configured pictograms, one set for all biological diagrams is infeasible with the approach taken. There are *mores* for the diagrams for each specialisation within the disciplines in biology and as long as you stay within your area, you'll be fine. Of course, this is impossible to do when you're in school because of the broad basis school teachers aim to impart, and so there's no other option to learn the various notations. That might become confusing. There's also an amount of guesswork in trying to read a diagram to piece together what its designer wants to communicate. Catley and Novick observed that at least for the cladograms, there's very little explanation in the examined textbooks about why a cladogram is the way it is. Where the source of that problem lies, remains to be seen.

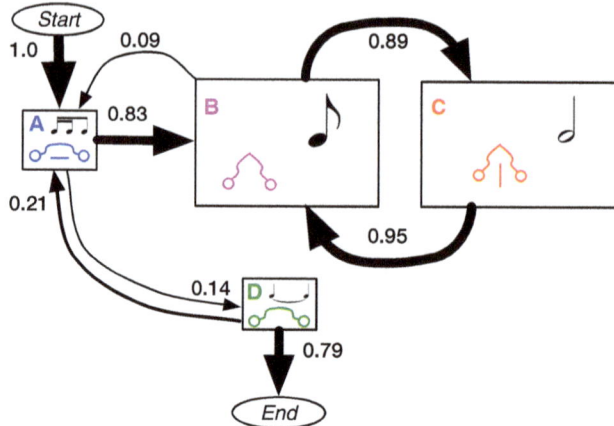

Relative Frequency of Occurrence and Sequential Relationship of Song Types during Dance Displays. Transition probabilities suggest the following typical sequence: from type A, they alternate between B and C several times before repeating type A and finishing with type D. B and C were performed at least four times more often than A and D, as indicated by box size. [...]. Transition probabilities less than 0.05 are not shown.

Fig. 3.7 A revised diagram of the superb lyrebird's show, which was tweaked thanks to the hints of the diagram design procedure

The biological diagrams have a limitation even if they were all drawn correctly: they are *pictures* that are hard to 'understand' by computers. Computers can process them to render them on screen, and show it interactively and with moving parts to simulate something, and a printer can print them, but it doesn't have a way to ascertain what's going on in the figure. Comparing pixel by pixel is doable, and that way an algorithm can recognise whether there's a cat in the picture, but no algorithm yet can figure out that two figures with the same content but a dissimilar layout, are *semantically* the same. This shortcoming is, arguably, not inherent to biological diagrams, but to the relative freeform permissiveness.

Drawing software may help achieving such automated processing of semantics to help determine if two diagrams model fully or partially the same information. Then we easily would be able to find overlap in images across scientific articles, find contradictions between them, and chart the refinements and extensions. We could check automatically whether they adhere to the rules they're supposed to adhere to and flag violations, thereby contributing to improving their quality. How could we do that? A drawing tool typically has a limited set of icons the modeller can use, such as an ellipse for kinases (a type of enzyme), a square for small molecules, and a dashed arrow for inhibition of some process, and so something can be gleaned from that together with the label attached to each shape. Then, if you're lucky, the diagram is saved in a structured text format, such as the XML or JSON standards for storing data on computers, rather than the mere image. Such formatted files

3.4 Limitations

are much easier to process, be it for the tool to convert it back into a colourful diagram or to compare one diagram to another such text-based representation that is at the back-end of the image. Generally, however, no one has access to those text-based files other than the creator and the company who developed the drawing software. The publishers of the scientific articles that the diagrams appear in often don't have access to the interoperable text-based files of the diagrams either. If, by a simple waving of a magic wand, publication policies would change and those files would need to be provided and made available as supplementary material upon publication of the article, and those diagrams are all neatly converted into plain text, and the software to process all the distinct formats of the structured plain text also instantly comes into existence, then we could start processing them henceforth as a new treasure trove of sorts.

Alas, not yet. Could it possibly be worth it to address this computational limitation of biological models? The answer to that is a resounding 'Yes!'. Kristina Hanspers, from the Institute of Data Science and Biotechnology in the USA, and her three collaborators wanted to know precisely that for the genes included in pathway diagrams. The written text of scientific articles is easy to search on the mentions of specific genes, but what is locked in the diagrams? To get there, they first queried PubMed Central, a large repository of scientific articles in biomedical and life sciences journals, for all images over a timespan of 25 years. Of those more than 200,000 figures, about 65,000 were characterised as molecular pathway figures following two rounds of machine learning. They were then subjected to an optical character recognition pipeline to find gene names. To make a longer story short: after all that processing and further analysis against scientific papers and databases, the discrepancies between diagrams and text was rather stark. They found, on average, 18.9 genes in a figure, but only 3.4 in the text of the article that has the figure, and overall, about a million mentions of genes in the figures but only about 100,000 in the text of those papers. Not just that, of the 13,464 human genes picked up from the optical character recognition, half didn't appear in the text at all and a little over a quarter of them weren't in the database checked (WikiPathways and Reactome). Or: computationally dredging the figures uncovered hitherto locked information missed elsewhere.[22] The picture *does* say more than the words. What else is locked in biological diagrams?

The reality check for the run-of-the-mill diagrams is that we haven't even started making the meaning explicit to be able to mine it in a less cumbersome way than Hanspers and her collaborators did. Time is implicit in the arrow-tipped lines. There are multiple meanings of the arrow-tipped lines, as we have seen already: to denote a chemical reaction, to denote movement, to denote who eats whom. The computer won't know if it hasn't been instructed somehow. Context is shown only partially, as we have seen with the cytoplasm and periplasm and the implications of that. Cardinality constraints, i.e., how many objects it involves or has, are missing, as are exclusions or disjointness to be able to specify, e.g., that zooplankton cannot be also phytoplankton. With that information absent by design, we won't know about any

---

[22] Details and further discussion can be found in Hanspers et al. (2021).

inconsistencies or contradictions in a diagram or across diagrams, like where one diagram puts the ACE receptor on the outside of the cell wall and another one put it on the inside of a cell. And, perhaps surprisingly considering all the nomenclature standards and unions and learned societies, there are no standards for the diagrams and the diagramming thereof in software. Not to mention the absence of any shown to be effective procedure for how to create them.

"Can a biologist fix a radio?", Yuri Lazebnik mockingly asked in 2002 when reflecting on research into programmed cell death (apoptosis), and his answer was a resounding "no". Engineering is highly modular, has lots of designs that can be compartmentalised, and processes are structured with methods and methodologies. Technicians can read the schematics for their modularised specialisation and fix it (if it is fixable). Lazebnik contrasted this with the certain lack of structure in cell biology and absence of a regimented modularisation and beyond-my-concern cordoning off from details. To be able to manage ever increased complexity, he opined, modularisation and abstraction are essential to get by. And so, if a biologist with their way of approaching biology uses that approach for investigating and fixing a radio (any electronic device, really), they would fail miserably. He called for taking some of that engineering approach to artefacts to enhance the biology approach to investigating objects and processes.[23] Interestingly, some biologists did that to some extent and we'll see more about that in Chap. 5. Let's look at some of that sort of engineering in the next two chapters.

References

Catley K, Novick L (2008) Seeing the wood for the trees: an analysis of evolutionary diagrams in biology textbooks. BioScience 58(10):976–987

Dalziell AH, Peters RA, Cockburn A, Dorland AD, Maisey AC, Magrath RD (2013) Dance choreography is coordinated with song repertoire in a complex avian display. Curr Biol 23(12):1132–1135

Driscoll CA, Clutton-Brock J, Kitchener AC, O'Brien SJ (2015) How house cats evolved. Sci Am Special Editions 24(3s):62–71

Gene Ontology Consortium (2000) Gene Ontology: tool for the unification of biology. Nat Genet 25:25–29

Halford B (2014) Reflections on chemdraw. Chem Eng News 92(33):26–27

Hanspers K, Riutta A, Summer-Kutmon M, Pico AR (2021) Pathway information extracted from 25 years of pathway figures. Genome Biol 21:273

Hu Yue OO, Karlson B, Charvet S, Andersson AF (2016) Diversity of pico- to mesoplankton along the 2000 km salinity gradient of the Baltic Sea. Front Microbiol 7:679

Keet CM (2005) Factors affecting ontology development in ecology. In: Ludäscher B, Raschid L (eds) Data Integration in the Life Sciences 2005 (DILS2005). LNBI, vol 3615. Springer, Berlin, pp 46–62

Keet CM, Berman S (2017) Determining the preferred representation of temporal constraints in conceptual models. In: Mayr H et al. (eds) 36th International Conference on Conceptual Modeling (ER'17). LNCS, vol 10650. Springer, Berlin, pp 437–450

Kritikou E (2007) How to build a biological model. Nat Rev Mol Cell Biol 8:424

---

[23] The entertaining and insightful article appeared as (Lazebnik 2002).

References

Lazebnik Y (2002) Can a biologist fix a radio?—or, what I learned while studying apoptosis. Cancer Cell 2:179–182

Lewis GN (1916) The atom and the molecule. J Am Chem Soc 38(4):762–785

Li G, Davis BW, Eizirik E, Murphy WJ (2016) Phylogenomic evidence for ancient hybridization in the genomes of living cats (Felidae). Genome Res 26(1):1–11

Lu Z, Zou Z, Zhang Y (2013) Application of mind maps and mind manager to improve students' competence in solving chemistry problems. Springer, Dordrecht, pp 235–245

Ostaszewski M, et al. (2021) COVID19 Disease Map, a computational knowledge repository of virus–host interaction mechanisms. Mol Syst Biol 17:e10387

Riehl JP (2010) Understanding chemical structure drawings. Wiley, pp 235–238

Rosse C, Mejino Jr JLV (2003) A reference ontology for biomedical informatics: the foundational model of anatomy. J Biomed Inform 36(6):478–500

Tett P, Wilson H (2000) From biogeochemical to ecological models of marine microplankton. J Marine Syst 25:431–446

Conceptual Data Models

4

> *So I think the key thing is to solve a real problem, to solve the problem people are worried and concerned about. And I just was lucky enough to solve at least one of those problems.*
>
> — *Peter Chen, inventor of conceptual data modelling and the ER language*

The biological models we have seen in the previous chapter allow for additional domain knowledge to be included in a diagram than is attainable with mind maps. More background knowledge is embedded in those biological models and thus also implicitly more information is represented in them compared to mind maps that are relatively superficial. Put differently, those models demand an increased exertion from our minds than mind maps require. But, for all they can do, those diagrams run into a wall as well. There are so many icons with implicit meanings that shift across sub-fields, a considerable number of people can't see the forest for the trees, nor, for that matter, discern the trees in the forest. Non-biology outsiders tend to find it confusing, it's easy to miss subtle distinctions without even realising it, and therewith it runs the risk of conceptual muddles or outright misunderstandings. One diagramming language to address this seemed infeasible.

And yet, that is precisely what computer science has done. They did not do that only for biological diagrams, but *regardless of the subject domain*. Software developers have to design systems for anyone—the scientists with their data from experiments; universities and companies to keep track of the data about their employees, students, and clients; governments to store data about the country's inhabitants; shops and factories to manage their inventory and sales, and even the neighbourhood's sports club membership database. There are people who have created their own database to manage their books and CDs collections and their baking recipes. Software developers do not start writing code out of the blue for such systems and hope working software comes out of it, just like architects don't design a house by putting stones on top of each other to see what comes out of it.

(Okay, there are coders who do, but they're not supposed to work like that.) There is a design phase beforehand. Models galore in that phase.

We did not devise a diagramming language for each and every domain and type of application. This is not to say that there aren't several types of models—there are, especially in the area of *domain-specific languages*[1]—but the key factor here concerns solving the relative 'mess' of proliferation of a wide range of diverse types of diagrams in biology. That can be solved. We do that by looking at the same stuff completely differently. We're going to shift the paradigm.

The crucial aspect is the notion of finding the commonalities across those models in different domains and devise a *modelling language* for those common things such that the language is independent of the subject domain. We don't bother with this sort of molecule or that sort of organism, or their interaction yonder, and an icon for each type of molecule. Those types of molecules are all *types of objects*. Those objects have properties themselves that are *attributed* to them, such as molecular weight or their size. The chemical reactions among the molecules make the molecules stand in a *relationship* to each other, as do the plankton relate to each other through a relation 'eat' or 'graze'. One can discern *roles* in those relationships, like that one participant plays the role of predator and the other of prey. And they relate to one another under certain *constraints*, like that each enzyme must have at least one active site where the action happens and that the 'married to' relationship is symmetric (e.g., if Romeo is married to Julia, then Julia is married to Romeo). Then, from those sort of common things—object types, relationships, attributes, roles, and constraints—we create a *modelling language*, which may be a graphical language. The models created with such a modelling language that we'll focus on in this chapter are called *conceptual data models*. These types of models and their languages weren't developed because of frustration with biological diagrams; it just so happens that they can be used for that too. Therefore, we're going to park biological models for a short while, look at computing's proposed solution that was trying to address a different problem, and then return to in what way conceptual data modelling solves those shortcomings of biological models.

4.1 What Is a Conceptual Data Model?

There are various types of models in computing, even among the information and knowledge-focussed models. A broad term for that is *conceptual model*. Within that group, there are again distinct types of models, especially in the realm of software engineering. The one that we are zooming into in this chapter are the conceptual *data* models. This is a class of models that capture the information about the data that are to be stored in the prospective software system. An example of such a model is shown in Fig. 4.1a about books, their authors, and publishers. We'll discuss what it all means in the next section. By way of informal comparison to the biological models we have seen the previous chapter, before reading the answer: can you

---

[1] See, among others, Reinhartz-Berger et al. (2013) and Fowler and Parsons (2010).

4.1 What Is a Conceptual Data Model?

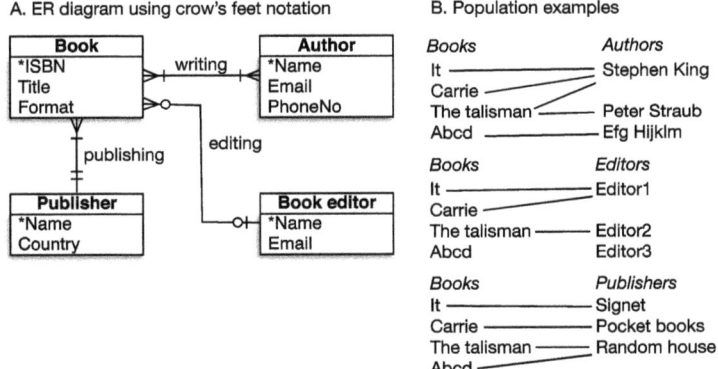

Fig. 4.1 A small conceptual data model in ER diagram notation with its crow's feet notation, about books, with on the right-hand side a sample population

describe what you think this diagram conveys and can you spot all the distinct elements used in the diagram?

4.1.1 The Beginnings—and Still Standing Strong: ER

Nowadays, massive amounts of data are generated—every mouse-click can be recorded, how fast you scroll on a webpage, and every interaction with a file store that you may have 'in the cloud' of Google, Dropbox, or WeTransfer, among many actions that are saved for data analysis. Before this Big Data era, smaller big amounts of data were generated, such as the recording of every transaction of every bank customer with their own bank and that of the one they received money from or transferred it into. Before that (and now still for that matter), there was what we now call 'small data': the employee database of an organisation, product sales of a company, and the VHS or DVD loans of the movie rental shop. Many computer scientists were scratching their heads in the 1960s already: how to store the data and how to access that data efficiently? Data management wasn't working well and various ways of storing, accessing, and querying data were explored by several researchers. Ted Codd was one of the computer scientists looking into that whilst working for IBM for most of his career. He proposed something called the relational model as a way of storing data.[2] It came with solid mathematical foundations. The relational model revolutionised data management and relational

---

[2] The original paper: Codd (1970). Edgar Frank Codd had an successful career well into his retirement in 1984, but it's the relational model that earned him an entry into the hall of fame of computing pioneers from the IEEE (see https://history.computer.org/pioneers/codd.html) (last accessed on 29-5-2023).

database management systems are still widely used today. It's not where the story ends, but starts, rather. It's 1970.

That relational model, while a demonstratively excellent way for data management, has quite a few nitty-gritty details that are relevant for implementing a database. That makes sense when you're heading toward an implementation, but if you're in the analysis phase trying to figure out what data needs to be stored and what their characteristics are, such implementation details and decisions are a nuisance. Implementation practicalities interfere with clear thinking required for the data analysis. That is, there's a need for a level of analysis and modelling a topic where we're not concerned with how exactly that data will be stored, but only what data needs to be stored. For instance, that the HR department of a company wants to record the special skills that employees have, not whether those skills have to be recorded in the same table as the employee table or in a separate table, or whether those skills have to be stored as a string of words separated by a comma for at most 255 characters or each skill be a word of at most 32 characters.

The solution for separating these concerns was proposed in 1976 by Peter Chen, who was also working for IBM at the time. He was a junior employee toiling away and trying to come to terms with that new thing for storing data, a relational database, which didn't go as smoothly as he would have liked. His solution set in motion a whole subfield within computing and information systems: conceptual data modelling. It earned him a spot in the hall of fame of computing pioneers as well. I had the honour of meeting him at the 32nd conference on Conceptual Modeling in Hong Kong, in 2013. He turned out to be a friendly emeritus professor who, to me, still seemed a bit confounded that the conference series was still alive and well, yet quietly relishing it as well.[3] The scientific conference on conceptual modelling is running yearly to this date, continuing to advance the knowledge, methods, tools, and specifications of conceptual data models.

What did Chen do to have become one of the pioneers in computing? He added another type of model, one that resides at one layer more abstract than the relational model: the entity-relationship (ER) diagram. It was dubbed to be a *conceptual* data model to indicate that it was positioned at the 'fluffy' layer that is independent of implementation—conceptual, not concrete, if you will, if you take computer programs to be concrete artefacts.

Instead of having to determine which data goes in which relational table and which tables there shall be, as if you were filling a spreadsheet with tables as you go along and add few advanced formulas that cross-reference between the tables, ER simply doesn't care about tables, because deciding on such tables are design

---

[3] It was not merely from afar in a large lecture hall or a mere passing in the hallway. Lunch was buffet style and there were round tables that could seat about 10 people. He sat at a table that was half empty and so I joined in. A (relatively) young academic, which I was at the time, is supposed to try to impress by showing off their smarts that they can hold their own with the academic VIPs. But they need a break, too, like everyone else. I can't even remember what we ended up talking about at the table, but we were cracking jokes and had fun about something that was only tangentially related to conceptual modelling.

and implementation decisions. Instead of the sprawling growth and intractable referential spaghetti, we're first going to determine what kind of things have to be recorded and processed. What we have, are entity types that relate to each other in some way. For instance, instead of declaring that each employee is identified by some ID, has a name, receives a salary, and works for a department in one relation, alike `Employee(ID, emp-name, salary, department)`, we can see there are two types of things, **Employee** and **Department**, where the former is a member of the latter, and the former has three properties: ID, name, and salary. That is what goes into the ER diagram. A sample ER diagram about books is shown in Fig. 4.1.

Does this difference matter? Yes, for if we were to learn that each employee can work for more than one department, the overall structure of the information in the ER diagram does not change. For the `Employee` relation in the relational model or the table in the relational database that would be disruptive, however, for it has the implicit assumption there would be only one department for each employee. It's the same story for the books and book editors of Fig. 4.1. To capture that many-to-many relation in the relational model, we need three relations: `Employee(ID, emp-name, salary)`, `Dept(name)`, and `EmplMemberDept(ID, name)`. In the ER diagram, we merely tweak the "–" end with a crow's foot. Having to revise a model substantially with even a minor update to the domain is annoying. An ER diagram helps clarifying this upfront, so that there won't be a need for extensive redesign of the relational schema and the database tables.

Mostly, the entity types in an ER diagram turn into relations in the relational model, hence the name Entity-Relationship. There are well-established algorithms for that transformation. For certain features in the ER diagram, there's more than one way to convert them into a relational model, providing further evidence that indeed those relational models embed lower level design decision and that conceptual data models, like ER diagrams, are entities in their own right.

What's in such an ER diagram? There are four key elements: entity types (drawn as rectangles), relationships (diamond shape or a line connecting the rectangles), attributes (ovals or another rectangle connected to the entity type rectangle), and constraints that adorn the relationship. There are a few distinguishable notations, but for ease of explanation, I'll introduce only two of them that are shown in Fig. 4.2— we don't know which of them is best anyway. As a lowest common denominator of all the ER diagram language flavours, a valid ER diagram is allowed to use only those core elements.[4]

That's a rather small list of elements compared to all the icons and arrows we've seen in biological models. The first key difference of these ER diagrams with the

---

[4] Arguably, that list also includes weak entity types, composite attributes, and multi-valued attributes. In the interest of readability, and because they are not present in all ER flavours and are used comparatively infrequently (Keet and Fillottrani 2015a), they have been omitted from this introductory overview.

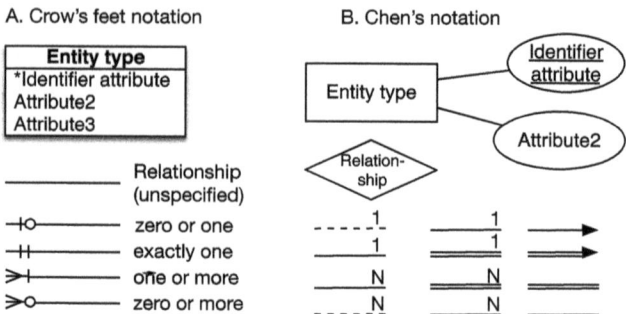

Fig. 4.2 Basic elements in the ER diagramming language. Left: ERD legend for those with crow's foot notation. Right: Notation by Chen

biological models is that there's a domain-independent 'vocabulary' of the things we can put in a diagram. The second key difference is that notion of constraints on the prospective data. It is possible to state that all objects of some entity type *must* participate in the relation (solid dot), or *may* do so but need not (white-filled circle). And state that one such object must or may be related to more than one object of the other type ('crow's feet' three little lines at the end), or just one (single line at the end of a line). Instead of just an arrow between an enzyme and a molecule, we now can state that one enzyme 'may bind at least one' molecule and, reconsidering the mind mapping example in Fig. 1.2 in Chap. 1, that an espresso machine 'must have exactly one' on/off button.

These basics have been augmented a little into *extended* ER, unimaginatively abbreviated as EER. The lines can be adorned with exact numbers, like a "1..5" for each book being written by at least 1 and at most 5 authors. And you're allowed to add subtypes, too. Subtyping is one of those things that's trivial only once you know it. Take, for instance, an entity type **Vehicle**: there are multiple types of vehicles. For instance, a subtype **Motorised vehicle**, which is a vehicle that has one additional property over just any vehicle: that of having an engine as part. A subtype of that again could be **Car**—which has yet one more characteristic: it has 3 or 4 wheels—and **Bus**—which as extra characteristic that it has at least, say, 17 seats. A sample diagram is shown in Fig. 4.3. One can pull the same trick for employees: an entity type **Employee** with as subtypes, for instance, **Manager** (those employees who manage a team) and **Area Expert** (those employees with a distinct skill and being a PhD graduate). And for books, instead of **Format** as an attribute, we can bump it up to an entity type and create three subtypes: **Hardcover** that has as extra property, say, the material, **Softcover**, with as attribute matte or shiny, and **ebook** with as additional attributes the platform and whether DRM applies. And that for any subject domain, really.

Is that it? Should you go design conceptual data models for databases with this information at hand? If only it were that easy. There are procedures for how to do this to help you along the way, which we shall see further below. Regardless of those

Fig. 4.3 Example with Extended ER: subtypes and further cardinality constraints, like the "4" identified wheels

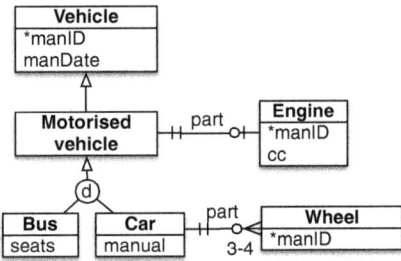

procedural aspects, is this small set of features enough to capture the knowledge shown in those biological models? Can we now model the mixed acid fermentation? Sort of, depending on how detailed you'd want it to be. Yes, easily even, if we are to model just that information from the diagram. No, if you'd want to explicitly model also implicit information as truthfully as possible. The features we need for making the implicit explicit, and thus actually moving the goal post, cannot all be done with this basic version of a conceptual data modelling language. Let's take a short excursion beyond the foundations.

4.1.2 Conceptual Data Modelling Explosion

As happens with every new solution in computing, there's a flurry of activity and a subsequent proliferation of adaptations, improvements, and alternatives in the early years that basically explore the potential solution space to try to find the optimal solution. It's needed to get to some resemblance of a stable state to become part of the canon. To give a few examples, there were tweaks in the diagrammatic notation that Chen proposed, such as only a line instead of a diamond shape for the relationship, drawing the cardinality constraints with craw's feet instead of only 1, N, and M, a dashed line versus a white circle for optional participation, small double lines instead of a black dot, and, in due time as monitors and graphics improved, the colouring in of the icons. Variations in modelling features of the modelling language was also explored, such as constraints on attributes. For instance, instead of assuming that each employee had 0 to any number of office phone numbers, one could now specify, e.g., 'at most 2' phone numbers. It's all fair game. Inevitably, at some point, there were attempts at standardising a notation, including IDEF1X and IE, and other trendsetters, like Barker notation and crow's foot notation. Standards are compromises. Those key developments happened before my time, and, from afar, it looks a lot like they weren't good in making compromises. Some tool developers, especially the catch-all ones, made it worse by mixing notations for their drawing canvases. It hurts.

The one clear agreement is the subtyping with separate lines, not as the cluttered Venn-like diagrams we had to draw when I was taking a course in databases over two decades ago. Roundtangles may look prettier, but rectangles are so much easier to draw on a canvas, so rectangles it is. As to the constraints, I have a strong preference for the crow's feet notation because it visually depicts the cardinality constraints. Have a look again at Fig. 4.1 on the right-hand side. Stephen King as one starting point has three outgoing lines, as if it's a 'crow's foot' notation already, and likewise from the single Talisman book with two outgoing lines to authors. The diagram stylises that informal visualisation as it were. Regarding drawing relations, I do have to admit that the notation with the diamonds for relationships is popular, among educators anyway. However, they take up much more space than just a line, so the diamond shapes may be sacrificed especially in scientific papers that have a page limit. Sometimes it's such externalities that determine notation, not any science behind it. It almost feels like I have to apologise for it, but we just have our priorities. Also, it's not trivial to investigate what the best notation would be for modelling a universe of discourse yourself and for understanding someone else's models. What we do know, is that notation affects what is modelled, and maybe then also how the modeller sees the universe of discourse. We don't have an answer to that now. Whichever notation you use: stick to one notation scheme throughout your model.

Be that as it may, in trying to use conceptual data modelling with ER, the modellers stumbled upon the fact that information had to be included in the diagram that was beyond what could be represented so far. Once you start looking at things in a certain way, you see more, probe further. One tweak concerned the attributes, so that mandatory/optional could be added as well as cardinality constraints. For instance, that Employee's attribute hasChildren is optional—if they have it, the number can be recorded and else it's simply ignored. It saves space on disk, which was an important consideration once upon a time. An Employee's taxNumber is likely to be mandatory and their officeNumber may be restricted from ranging between 0 offices for flexworkers and 2 for employees housed in two departments. They're all plausible and the more such constraints are captured in the EER diagram, the more constraints can be implemented in the database and the cleaner your data will be and thus of a better quality. That results in better business analytics and pretty dashboards that show improved truthful information for better management in the organisation. In theory anyway; good data won't fix a toxic or clueless manager using that data.

With that, we have climbed the ER mountain. These foundations are still taught in undergraduate degrees in computer science, information systems, and IT. The view from this top not only shows the valley whence we came from, but also another mountain that lies behind it—after all, between ER notations settling in the late 1990s and now, there has been over 25 years conceptual modelling research and experience. We're not going to climb those new peaks here, but only touch upon what those peaks are made of. It's full of orthogonal extensions. What we can do so far with EER, i.e., model in our conceptual model, is confined to static information about objects, their attributes, relations, and constraints. What EER can't do is state things related to time or space, or things that aren't crisp. For instance, a company may demand rising through the ranks to grasp what's happening in the organisation and so any Manager thus *must have been* just an employee (i.e., not managing) *before* they can become a manager, or that a river is a *spatial object* of

4.1 What Is a Conceptual Data Model? 57

type oriented line and the area of a city is enclosed in the area that makes up a country. Unlike mind maps, where we can't change anything, and biological models with its unchecked freedoms, with conceptual data models, we can add extensions in a controlled manner. We're going to do that now to illustrate the idea (rather than introduce a whole new language, about which one can write a separate book), and along with it, fulfil the promise that we'd revisit the mixed acid fermentation in the context of conceptual data modelling.

4.1.2.1 A Temporal Extension to ER

Many chemical reactions occur during the mixed acid fermentation, where one reaction happens *after* the other, or, more precisely: if a later one in the sequence is happening, then the one(s) before that must have happened. How can we represent that in an EER diagram? With plain EER, the best we can do is to represent the objects and how they relate, and fake the before/after with a relation that we could name, say, reactionOccurredBefore for relating the chemical reactions and a producedFromPrecedingMolecule for relating the molecules. The words make sense to humans when they read a diagram, but not to the computer. The computer merely sees a relationship labelled with a string consisting of 22 (respectively 29) letters. A natural language processing algorithm can figure out there are three words in those 22 letters (four in the 29), that it is a noun phrase, and that 'reaction' is a noun, 'occurred' is the verb, and 'before' is the adverb (a likewise process for the other). That's all. To enforce the notion on the data so that it will be stored correctly in the database, we need new constraints. Constraints that deal with *time*. There are many ways to do this at many places, but at the conceptual modelling layer where we are now, we can add a diagrammatic notation for that to EER, such as with the ER_{VT} language or its successor TREND.[5] The 'ER' is our usual ER diagram and '$_{VT}$' stands for Valid Time: the time as it is in reality, which contrasts with 'transaction time', which is when something changed in the database recording the thing or event. Transaction time is relevant for storing Bitcoin data where time stamps as recorded in the system. ER_{VT} was unavoidable during my PhD and stint as assistant professor at the Free University of Bozen-Bolzano in Italy: the co-supervisor of my PhD, Alessandro Artale, was integrally involved in designing ER_{VT}.

We used it in several research advances, but even so it felt like still more should be possible with the whole machinery. ER_{VT} has temporal entity types for objects that at some point in time are an instance of another entity type and Alessandro and I had extended the idea to temporal relations. The third key element of ER diagrams is attributes, so perhaps temporal attributes may be relevant, too. This is what one of my students, Nasubo Ongoma, investigated for her master's thesis. Conclusion: indeed, they may be relevant and so they were added as well. The new modelling

---

[5] For earlier logic-based approaches and a solid literature review about the key aspects of temporal conceptual data modelling and ER_{VT}, refer to (Artale et al. 2007). The latest and most comprehensive temporal EER modelling language, called TREND, is described in (Keet and Berman 2017).

language was unceremoniously dubbed ER_{VT}^{++}. Then there was the question of notation in the ER_{VT} and ER_{VT}^{++} diagrams: Alessandro and his collaborators had made it up on the fly to have something to offer, as did we for our "++" extension, but is it any good for drawing and understanding temporal ER? That turned into another dissertation by one of my masters in IT students, Tamindran Shunmugam. In the same line of thinking: maybe a controlled natural language to switch from diagrams to structured text may help the modelling as well. That design and evaluation, I did. Then there were further extensions and pondering about notation with my colleague Sonia Berman, also with the University of Cape Town, which resulted in the latest and most comprehensive language around in the world, which we deftly called "Temporal information Representation in Entity-Relationship Diagrams" to make it abbreviate to TREND. We've evaluated it extensively in experiments on our students who learn about ER and fine-tuned the notation of it, since they still preferred diagrams over text. Meanwhile, we're some 15 years hence. Things go slow, but progress is made.[6]

What feature is available in TREND that is needed for the mixed acid fermentation? Since conceptual data modelling is independent of the subject domain, we're not going to add those clunky long names to the model itself, but also here we'll take an approach that is reusable across subject domains. Ethanol *evolved from* acetaldehyde, for instance, as did the butterfly evolve from the caterpillar, and an employee may *also become* a manager. Slightly more precisely: if we have an individual butterfly at time point t1, then there must have been an individual caterpillar at time point t0, where $t0 < t1$. This can be formalised in such a way that both the computer can process it to check whether it's consistent and for humans to understand it. If you happen to know some logic, here's the precise meaning of the constraint for objects migrating from being an instance of one entity type into also another, where the DEX indicates 'Dynamic extension in the future', like the employee also becoming a manager:

$$o \in \text{DEX}_{C_1,C_2}^{I(t)} \rightarrow (o \in C_1^{I(t)} \land o \notin C_2^{I(t)} \land o \in C_2^{I(t+1)})$$

For the non-logician humans reading and drawing TREND diagrams, a new element is added to the set of permissible icons: an arrow. The new constraint can be used both for objects that evolve from one thing into another and for things that are an instance of one entity type and then also become an instance of another entity type, like a student also becoming a teaching assistant. Some objects must do this, others have the option to do so, which is indicated with a solid and a dashed arrow, respectively. If it was in the past, there's a "−" superscript. For instance, the butterfly relates to the caterpillar with a DEV− temporal constraint for Dynamic EVolution: it must have happened in the past. The '(may) also become' is indicated with DEX,

---

[6] The extensions and updates to ER_{VT} are described in, respectively: Keet and Artale (2010), Keet and Ongoma (2015), Ongoma (2015), Shunmugam (2016), Keet (2017), and Keet and Berman (2017).

4.1 What Is a Conceptual Data Model?

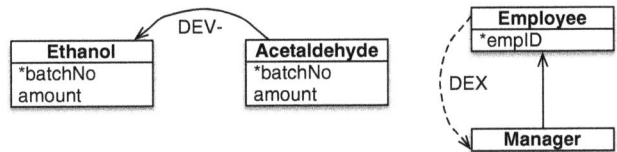

Fig. 4.4 Two examples of representing temporal information: some amount of ethanol was produced from acetaldehyde and is not ethanol anymore (DEV⁻ with solid line), and an employee may also become a manager (DEX, optionality indicated with dashed line)

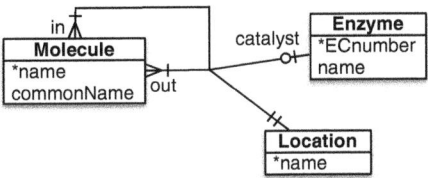

Fig. 4.5 A way to model the mixed acid fermentation in a conceptual data model for an atemporal database, with a quaternary (4-ary or 4-way) relationship in which Molecule participates twice

an abbreviation of Dynamic EXtension. A few examples are depicted in Fig. 4.4, including an explicit one for ethanol under the assumption it's a conceptual model for a factory producing the chemicals.

We can do likewise for other temporal constraints as well. For instance, for relationships; e.g., a divorce between two people should be allowed to be recorded only if there was a record of marriage between them before. Or to 'freeze' an attribute, that once the value is set, it cannot be changed anymore; e.g., if a marriage is registered, it happened on a certain date with a certain ceremony person, which are facts that will never change in the future.

There are also 'workarounds', or ways, to model information that is inherently temporal, atemporally. Attributes for recording time may help, or to just assume the feature to change, like it's implicit in the biological models. The generic diagram for those enzymatic processes may then look like the one shown in Fig. 4.5. Practically, what the model will be depends on what needs to be stored in the database as to which options is most suitable. It didn't use to be deemed relevant for databases, but goalposts shift. Business process modelling is popular at the moment. There are objects participating in all those processes. Objects that need to be stored in databases. Objects that need to adhere to constraints coming from that business process logic. Temporal conceptual data modelling can cater for that.

4.1.2.2 Yet More Extensions

Examples analogous to the temporal one have been devised to motivate for the other extensions, being mainly for spatial and fuzzy data. Fuzzy is a type of uncertainty and may be used when we can't or don't want to figure out crisp boundaries for entity types. An object then be an instance of an entity type to a certain degree.

For instance, this book is decidedly a book, but what if it were to have had only 50 pages? Would it then have been a book still, or rather a booklet or a pamphlet or a lengthy article? And what if it's 35 or 75 pages? Fuzzy extensions can process this, but I'm not sure to what extent it is used in industry.

The most marketable and widely used extension concerns spatial databases and it's a core component of geographic information systems (GIS). GIS are nontrivial and they are so widely used that there are separate departments and degrees at universities. When you use Google Maps or OpenStreetMap, you consult a GIS database; when you go skiing in Switzerland and check for avalanche danger beforehand, you query a GIS database; cadastre information for buying a house; where exactly all the public transport is in a city; mapping flooding danger for coastal zones due to climate change; movement of whales in the oceans; forests and forest fires; the Environmental Data Explorer of the United Nations Environment Programme. There's also a conceptual data modelling language were both space and time were added, and, I suppose, one could fuzzify that.[7] And summarise and modularise such a model, too. And if we already had the next two chapters under our belt, expand it further. We have enough for our purposes, though, so we'll leave all that be.

4.1.3 On Turf Wars and Truces

EER is not the only player in town. Computer scientists saw ER and some of them thought 'we can re-purpose that notion for our own good!', as did others in varied ways, and yet others zoomed in on notational variants or adding and tweaking constraints to create similar conceptual data modelling languages. It's a typical process once a new successful technique has been proposed. Conceptual data modelling languages proliferated in the 1980s and early 1990s, to eventually settle on three main families of languages: the EER that grew out of the relational databases (often still called ER, saving space and typing one letter less), UML for object-oriented programming, and Object-Role Modeling (ORM, originally called NIAM, and also known as fact-based modelling) that can cater for both purposes and for business rules. Looking back at it now, from a place where we have unified the languages, part of the fights seems a petty waste, but there are both obvious and subtle facts and arguments that otherwise may not have emerged and therewith increased insights. I'll introduce them briefly.

4.1.3.1 The Unified Modeling Language

UML stands for Unified Modelling Language and is a standard of the Object Management Group since 1997. When I learned it at university, in UML version

---

[7] A good example of a conceptual data modelling language with a spatial extension is MADS, which was motivated by a use case about avalanche prevention for the Swiss Alps (Parent et al. 2006).

Fig. 4.6 ER example of Fig. 4.1 in UML notation. It uses so-called 'look-across' notation for the constraints: the multiplicity constraint is at the other end of the line

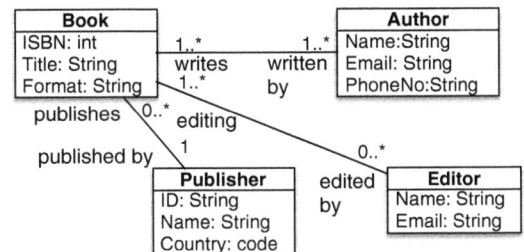

1.4 times, they covered, if I remember correctly, seven disparate types of models. Now, with UML 2.5.1 that was released in 2017, there are 14 types of UML models that are described in 796 pages. Of those, the UML Class Diagram is the object-oriented programming version of ER. The purpose is still the same: a model of the data to be dealt with in the software system. The UML class diagram version of Fig. 4.1 is shown in Fig. 4.6.

The UML class diagram is normally not translated to a relational database, but to object-oriented code for computer programs, with object-orientation being one of the programming paradigms. That aim downstream does affect the model upstream measurably for the publicly available models that I analysed with my collaborator Pablo Fillottrani of Universidad Nacional Del Sur, Argentina. We had our suspicions from experience and anecdotes, but the plural of anecdote is not data. The tedious task of tabulating the language features used in 101 diagrams paid off with insights based on evidence. It turned out that there were three key differences between UML Class Diagrams and ER, besides naming a subset of the elements differently.[8]

First, there are *methods* listed in UML class diagram that are the respective abstractions of the things to be done, because in object-orientation there are objects doing things to other objects. ER doesn't. Those methods are part of the specification of the UML class (equivalent to the ER entity type) that's doing it. Second, in programming, there are clean-up operations of orphan objects, and it would be nice to have to do less of that manually or in a better controlled way. Enter the *aggregation association* in UML class diagrams, which is a special type of relationship, called an *association* in UML parlance. It allows the modeller to specify that if the whole is destroyed, so must all its parts be deleted. For instance, if the soccer team Manchester United is disbanded, then all its members, as members of that team, will be removed from the software system as well. And since there's that special element for such types of relation, it's being used. Not only that, the aggregation association is used more often than the relationships in the assessed ER diagrams that have roughly the same meaning with labels such as 'member', 'part' or 'component'. The suspicion is that because it's there as a language feature in UML, enough modellers are curious and 'see' a larger number of those relations

---

[8] The UML standard can be accessed at https://www.omg.org/spec/UML/ (last accessed on 29-5-2023). The article about the observed distinct characteristic features of ER diagrams, UML Class Diagrams, and ORM models is Keet and Fillottrani (2015a).

in the subject domains and therefore use it more frequently. Third, UML models have a larger number of *hierarchies of classes* that are deeper. While no-one has investigated why this is so, it is probably for two reasons. ER emphasises relational aspects of data and in its first incarnation, as ER, it did not even have subtypes as a language feature. In contrast, class hierarchies were emphasised from the start in object-orientation, as it nurtures a programmer's laziness in favour of reuse, because for each subclass, only the new or modified attributes with respect to its parent class have to be added, not all of them. It's still the same notion of subtyping as we've seen with EER, but then called 'inheritance' of the features.

4.1.3.2 The Object-Role Modeling Language

There are big companies that are heavily involved with the UML standardisation as well as many modelling tools. Unsurprisingly, UML is popular. Still, ER survives to this day. There's a third family of conceptual data modelling languages that is not as well-resourced as UML and, like EER, persists and survives since the late 1970s: the Object-Role Modeling (ORM) language. They're the best of all of them for structural conceptual data modelling for both object-oriented programs and databases and also for declaring business rules in a systematic way. Well, that's the claim anyway.[9]

ORM certainly has multiple distinct advantages—with sound argumentation, not just the pedestrian look-here vs. look-across debate on where to place a constraint in the diagram—and therefore deserves to be mentioned as well. In addition, ORM 2 has 32 types of constraints, compared to UML class diagram's 16 (counted generously) and EER's 13 and so more information of the business logic can be captured more precisely.[10] Also, ORM originates from a distinct angle towards conceptual modelling. I had stumbled upon ORM in 1998 when I posed as a student at the University of Applied Sciences in Enschede in the Netherlands and bought cheap printed and bound lecture notes of some of its courses of the informatics programme to augment the conversion course I was enrolled in nearby. The bound notes and ORM resurfaced—by vague recollection and digging into, and dusting off, my fine collection of notes—during my honours database project in 2003. ORM enabled teasing more detail out of my client and thanks to that I developed a better model and therewith a better database. I became an ORM enthusiast and therewith unsuspectingly walked into the endless turf wars of the 'my preferred language is better than yours' variety before I even knew there could be such heated debates about them. That was in 2005, in my first year of PhD studies. I was curious to meet other ORM enthusiasts and a visit to Cyprus, where an ORM workshop was to be held as part of a set of conferences and workshops, also sounded very appealing. I did the research involving ORM, got the paper accepted, and my supervisor had the money to pay for me to go. The first morning of the workshop had an old arrogant man sitting within earshot from where I was. He had popped in to hear

---

[9] The first definitive ORM book is written by Halpin (2001). It contains extensive comparisons with UML and ER and EER and arguments why it is better.

[10] See Keet and Fillottrani (2015b) for details on which language has which constraints.

4.1 What Is a Conceptual Data Model?

what ORM was about, but at the same time being very convinced his UML was naturally superior. I started countering some of his utterances with arguments. He made it crystal clear that, in his opinion, surely a young woman couldn't possibly know what she was talking about, and on top of that to dare to criticise his beloved UML that he had worked with for years. Unacceptable. And yet, he capitulated and became a convert, because the same arguments out of the mouths of fellow old white men were convincing eventually. We never got along as I was already firmly put into the 'bad person' box; scientists are human, too. Him and people like him make researchers reroute their 5–10-year research plans. I'm glad I didn't have to live through the worst days of the turf wars. ORM is cool and I wouldn't want to have to give it up permanently.

ORM has its origins in semantic networks and natural language. The first version was invented by Sjir Nijssen, then at Control Data in Brussels, Belgium. It was called the Nijssen's Information Analysis Methodology, NIAM, since it combined the notation with a way to go about designing such a diagram. Later, it was re-branded as Natural language Information Analysis Methodology since an increasing number of people was involved in shaping it all, and eventually into CogNIAM, with 'cog' for 'cognition enhanced'. I've also met Sjir Nijssen, at one of the ORM workshops in the 2000s. He's engaging and passionate about ORM, has left academia a long time ago, and set up a successful consulting business, PNA Group, that his son Maurice meanwhile has taken over. They're still using CogNIAM successfully in their projects with government and industry. Then came Terry Halpin, who liked NIAM but wanted a logic-based foundation to it, which he did, and it earned him his PhD. He has had a defining influence on ORM since then, both for academia and in industry when he worked for Microsoft—they bought up the modelling tool he designed—and, later, LogicBlox. Unlike Nijssen's jubilant enthusiasm of the 1970s pioneers, Halpin came of age academically during the heights of the modelling languages turf wars era, and it shows. We've met several times at ORM workshops, and it was inspiring time and again. Besides being a prolific writer of very readable and comprehensive books and an avid tool designer, he also can provide detailed analyses and modelling suggestions when put on the spot, and he knows all the arguments of the comparisons by heart. I had the distinct impression that he found it still a bitter pill to swallow that it's not always the scientifically best option that makes it into mainstream.

Notwithstanding, and thanks to contributions from a waxing and waning community of aficionados, meanwhile ORM has its own extensions and such, analogous to ER. There are several logic-based reconstructions, notational variants (e.g., FCO-IM), modelling tools, and language extension. What sets it apart from ER and UML, is that its notation is used for other tasks in computing as well, such as being the foundation of the Semantics Of Business Vocabulary And Business Rules standard of the OMG. That same OMG of the UML standard.[11] A sample ORM diagram

---

[11] Halpin's PhD thesis is available as a pdf online after he found a student to scan it (Halpin 1989). There's also a 2nd edition of the ORM book, by Halpin and Morgan (2008). FCO-IM (Fully Communication-Oriented Information Modelling) is described by (Bakema et al. 2005) in Dutch. Among the tools, the NORMA plugin to Microsoft's Visual Studio is recommended, which is available open source at https://github.com/ormsolutions/NORMA (last accessed on 29-5-2023).

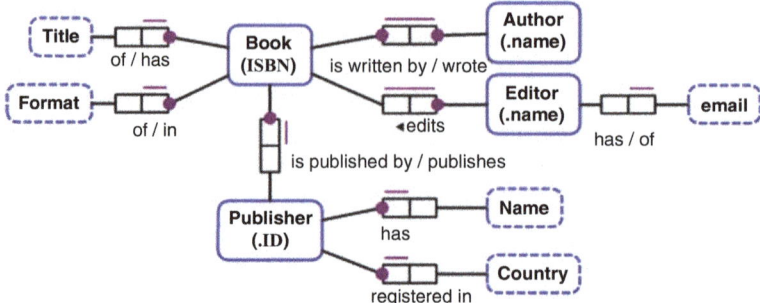

Fig. 4.7 ORM notation of the ER example of Fig. 4.1. The constraints apply to the side of the fact type they are declared

drawn with NORMA is included in Fig. 4.7, which conveys the same information as Fig. 4.1.

Some key differences in the ORM language 'package' compared to UML and EER is, first, that ORM is a package: the diagram notation, (pseudo-)natural language rendering of the diagram, and the sample data approach. Second, ORM is so-called 'attribute-free', unlike UML class diagrams and EER. ORM represents the object types and the value types (otherwise attributes) all the same level, participating in a fact type, rather than one a class and the other, secondary, as attribute of the class. This can be seen in Fig. 4.1 with the roundtangles that have a solid line versus a dashed line; e.g., compare Editor to email. A consequence is also that the ORM model is more stable to changes in the subject domain semantics or what needs to be represented, because it embeds fewer design and implementation decisions. It makes the conceptual data modelling language more conceptual than the other conceptual data modelling languages. Holier than thou, or 'more Catholic than the Pope himself' in Dutch idiom. Third, there is a much larger number of types of constraints in ORM compared to UML or EER. Indeed, only a few are used most of the time, but advanced modelling instances do exist and ORM has got it covered, and those esoteric constraints are used somewhere. We discovered that in the same assessment of models as the one mentioned earlier on UML. For instance, on two persons marrying each other, you could add a constraint that relation is irreflexive so that a database will ensure that no data will be added that a person were to be marrying themselves.

4.1.3.3 Compare and Combine

Three distinct origins with distinguishable purposes in designing software—relational databases, object-oriented programming, and both plus requirements engineering—resulted in three key conceptual modelling languages. They can't

---

The SBVR standard is accessible from https://www.omg.org/spec/SBVR/ (last accessed on 29-5-2023).

easily be swapped. Squeezing UML's methods into ORM or ER is awkward at best. ER has lots of relationships that involve more than two entity types, which is unpleasant and confirmed to be confusing in UML notation.[12] ORM diagrams take up more space than the other two and the abundance of constraints seems to confuse the inexperienced modeller and requires more detail than one may care about. By now, people have accepted they have their respective user base for their own reasons, and each modelling language has its advantages and disadvantages.

My collaborator Pablo Fillottrani and I accepted that several years ago and have been working on trying to make the main conceptual data modelling languages compatible through a common metamodel, transformation rules, a common core as logic foundation with three profiles, process workflows in an interoperability framework dubbed FaCIL ('easy' in Spanish), and even a tool (called crowd2.0). It allows one modeller to model in one language, and another modeller to model in another language, yet at the back-end behind the scenes, they map into the same instantiation of the metamodel for as much as possible. Any conversions between models in the supported languages are computed automatically.[13] Scientifically, it's sound, but whether it's going to gain traction remains to be seen. At least the technology now allows for, say, a database designer and EER enthusiast to collaborate smoothly with an object-oriented UML user and an ORM aficionado. A truce it has become, or so far as possible: one concretely can choose one's preferred language and work together in one application.

4.2 How to Develop a Conceptual Data Model

There are several established procedures and academic proposals for how to develop a conceptual data model. The question of what the *best* way to develop one would be, may not be answerable in this broad formulation. The question can be narrowed down to what works for a majority of modellers for a particular conceptual data modelling language, which is answerable in theory. I am not aware of any research that has investigated that by means of method comparisons. Notwithstanding, insights into methodological approaches for the development of conceptual data models are many miles ahead of those for the biological models that we saw in the previous chapter. And you even can *choose* among procedures!

While research into conceptual data modelling languages is firmly within the realm of computer science, the processes around the development of the models has one foot in computing and the other in information systems, and with the weight on the latter by scope. The well-known procedures proposed, however, come from the

---

[12] (Shoval and Shiran 1997).

[13] The key references are: the metamodel (Keet and Fillottrani 2015b), the metamodel-driven approach with the rules (Fillottrani and Keet 2014), the evidence-based logic profiles (Fillottrani and Keet 2021), and the FaCIL workflows and implementation in the crowd2.0 tool (Braun et al. 2023).

computing side, to give some general guidance. The basic procedure for especially ER diagrams that is the staple taught at universities, is as follows, in a nutshell:

1. There is some text that is a natural language description of the data requirements for the system to be developed. It comes from another process elsewhere that is of not our concern. This text is very nicely worded for our purpose.
2. Read the text in its entirety.
3. Read it again, but now underline the key nouns that may be candidates for entity types. Candidates are those things for which data will be gathered and need to be stored; e.g., "books" for a library loan database.
4. Read it again, and squiggle-underline the key verbs that would relate those nouns; e.g., "borrows" for a library loan database.
5. Read it again, and dash-underline the key features of those entities referred to with nouns and verbs; e.g., "title" of a book for a library loan database.
6. Assess the three types of underlined words and make a list of the vocabulary to settle on, removing synonyms; e.g., if there was "name of the book" and "title" dash-underlined, which refer to the same kind of thing, then choose one term for it.
7. Create the diagram. This involves: draw entity types from the relevant key nouns, attributes from the features of the entities represented with the nouns, and relationships from the verbs between those entity types, and adding the constraints (cardinality, disjointness etc.).

This procedure functions as a first pass to help students in computing to develop those models. It's based on educational principles rather than whether it would be the best approach scientifically to draw a conceptual data model once known: they—as did I during my studies—start with something they know, being natural language text, nouns, verbs, attributes, and they are taken by the hand to create something new that they're learning about. Here's an example of such a paragraph of text for a prospective database about employees of a company:

> Kathy's clothes emporium has been hiring personnel and needs a better employee system. There are three types of employees: designers, salesclerks, and administrators, who must work in teams of more than one employee. They are identified by their ID number, but also name, telephone number and address should be recorded, as well as their salary. Designers must have a diploma. Teams are identified by their name. Their creation and disbandment dates must be recorded as well. Any employee can be a member of at most 3 teams.

After those steps of annotation, the text may look like this:

> Kathy's clothes emporium has been hiring personnel and needs a better employee system. There are three types of _employees_: _designers_, _salesclerks_, and _administrators_, who must work in teams of more than one employee. They are identified by their ID number, but also name, telephone number and address should be recorded, as well as their salary. Designers must have a diploma. Teams are identified by their name. Their creation and disbandment dates must be recorded as well. Any employee can be a member of at most 3 teams.

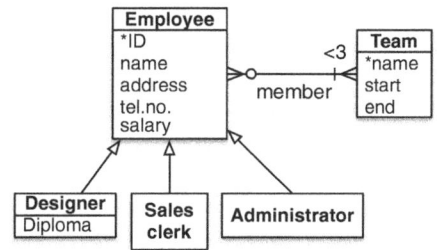

Fig. 4.8 ER diagram of the example with Kathy's clothes emporium

The resultant model is shown in Fig. 4.8.

If it seems a bit of a leap from the text to the diagram: it is. Normally, making people acquainted with this skill happens during in-person lectures and group assignments. Written text isn't the most effective means to communicate the intricacies of the processes, just like the drawing process of the biological models and mind maps hasn't been spelled out in the literature and is expected to happen during the in-person education. In addition, the steps listed above are typical, but not completely in cookbook style, or, more precisely, cake baking style since baking is stricter than cooking and some level of precision is asked for in the process. Textbooks may have further details, but we know there are two well-known gaps, or blind spots, in this 7-step procedure. First, the "create the diagram" step is sketchy on how that is supposed to happen. Second, it rests on the assumption that there indeed is text and that one would start with text. There may well be actual data already but not text, or there may be lots of text but no neat succinct and to-the-point text that describes the domain, or there's only a sample of what sort of data should be stored in a database. There are people who had noticed the gaps as well and have been working on addressing those. Whether such a proper design process is taught at university or technical college largely depends on the degree or diploma programme structure and time available for such matters. I'm including one in the next section to make the point that there is such a thing as a design procedure for the models themselves. If it raises appreciation for the task, and sticks, all the better.

4.2.1 A Conceptual Schema Design Procedure

A concrete example of a methodology that goes into the nitty-gritty detail of the actual authoring (drawing) of the diagram, that is, that "step 7" of the general procedure, is the Conceptual Schema Design Procedure. The procedure is designed for ORM, but the principles can be adapted for other conceptual data modelling languages just as well. The procedure is described in detail in Terry Halpin's book, with 'detail' taking up 242 pages in a tome of 949 pages.[14] We shall not rehash

---

[14] It's described in (Halpin 2001) for ORM and in (Halpin and Morgan 2008) for ORM2 that has an updated notation and a few additional constraints compared to ORM.

those details and double the size of this book. My aim here is to humbly plant a seed for diagram authoring methods and methodologies: it *is* possible to have such systematic processes for designing the models themselves. It neither has to be relegated to an art form, nor may only expert wizards call a model into existence by the mere utterance of abracadabra. A more fitting analogy would be that of building a house or a bridge—an exercise in engineering, in developing an artefact based on design principles, theory, processes, and heuristics. The Conceptual Schema Design Procedure has seven steps as well, to which I have added brief illustrations as we go along.

1. Transform information examples for your application domain into so-called *elementary facts*, which are as short and simple as possible without losing information, and apply quality checks that the data is correct. Those facts may be a table or a spreadsheet with data or already basic phrases with such data. For instance, and pedantically, "The *author* with ID 12345 *wrote* the *book* with *title* 'Hamlet' ".
2. Draw the fact types—that is, the classes with their relationships—and populate it with some of that sample data. Continuing with the example, there's an object type with identifier, formulated as a basic fact type **Author is identified by ID**, and a fact type between object types as **Author wrote Book**, and sample data for the are (12345, 'Hamlet') and (56789, 'No taming of the enthusiast').
3. Double-check for the possibility where entity types have to be merged because they're the same or they may have a common parent type, and take note of any arithmetic functions (derivations in ORM parlance) that something of interest can be calculated.
4. Add basic 'uniqueness constraints' to each fact type. These constraints determine the cardinality, or: how many of one of the role players can link to the other role player(s) in the fact type. For our book example, let's assume that a book may be written by more than one author and one author may write more than one book. So, to get unique rows in our table, we need to take the combination of the author and the book and draw a line over the whole rectangle shape of the fact type. (If there were a land where books can be written only ever by one sole author and they only ever write one book, then there's a 1:1 relation, with a uniqueness constraint over each part of the rectangle.)
Also check the arity of each fact type, i.e., how many entity types participate in it, and whether it can be split up into multiple fact types without losing information. This relates to step 1 insofar as a fact types may fall through the cracks of not having been recognised as not 'elementary'. A fact type is elementary if the uniqueness constraint spans all n, or n-1 roles. For instance, a fact type of "Author wrote Book and participated in Book launch" can safely be split up into two fact types, without losing information: 'Author wrote Book' and 'Author participated in Book launch'. It is a different story for "Author participated in book launch of the book they wrote." and "Author wrote book that was launched on date x", which both *are* ternary fact types.

4.2 How to Develop a Conceptual Data Model

5. Add mandatory participation constraints of the entity types in the roles, where applicable. For instance, at least one author must have written that book (even if that author is an algorithm). Also, check for any logical derivations, which is when one fact type can be derived from another. For instance, if the Book category determines whether a Book review will be commissioned, and each Author writes books in a Book category, then we can determine whether the author will get their book review.
6. Then deal with the first group of more advanced constraints: value types with their constraints. For instance, a published value type may have only a {'Y','N'} as permissible values, i.e., either it's published or it's not. The 'basic more advanced' constraints include set comparisons among the fact types, and subtyping constraints on entity types, like that romance novels are books.
7. Finally, add any other applicable even more advanced constraints, like reflexivity or acyclicity, and perform final checks on everything that all is in order and does indeed cover all those information needs from the preceding requirements engineering stage and from step 1 have been met. For instance, if authors are only allowed to support other authors but not themselves, a fact type "Author supports Author" may be adorned with the irreflexivity constraint to ensure that a fact like supports(12345, 12345) is never recorded in the database.

I've tried it and it works. Especially when I started out with conceptual modelling, having this procedure at hand was useful. It's a task that seems to be missing from the ER canon: instead of willing the diagram into existence by sheer force of mind, which no human can do, this procedure easily can be adapted to the ER context, and even UML, so that anyone can take it one step at a time to get there.

Intertwined with the turf war debates on different conceptual data modelling languages, at one of the ORM workshops, the question 'how do *you* model?' was posed. The participants ended up split in two, with a passion. There were three key categories for one aspect and two for the other. First, what we start with: (1) the natural language analysis of the sentences for the fact types, (2) drawing the diagram, or (3) the sample instances. Second, the sequence of adding content to the model: (1) put down all the elements first and then add the constraints, or (2) alternate between elements and constraints. The ones who preferred the former thought along the lines of 'get all your elements in there, even if with the weakest constraints, so as to not miss data' and the second one 'whatever data there has to be, make sure to get it precise and right for what you have' even if you miss some of it that has to be added later. Diverging priorities, with no end in sight that a consensus might be reached. There was at the time, and to the best of my knowledge still now, no evidence whose theoretical arguments and assumptions win in practice. To resolve that, it has to be put to the test scientifically with a controlled experiment. A known unknown that nobody cared enough about to investigate. For the first category of key difference, on starting with natural language, diagramming, or examples: we know it depends on the stakeholders' and modellers' background and preferences and each approach can be motivated as a sensible one from a cognitive science perspective. The most popular, if not all, ORM tools allow for each of the three starting points, so you can pick what works best in the circumstance.

4.2.2 Top-Down and Bottom-Up Approaches

Besides a complete *procedure* like the Conceptual Schema Design Procedure, there are what one may call 'approaches': the angle from which you start developing the model. In theory, it's probably valid to say there are top-down, bottom-up, and middle-out approaches.

Top-down approaches assume there are organising principles handed down from the experts among the experts that the rest of the analysts, architects, and modellers will follow and possibly adapt a little. Senior experts have noted commonalities across organisations and documented modelling solutions of recurring challenging data needs. The generic model snippets that are common throughout certain organisations are part of a so-called *enterprise model* or of an *ontology*, which is, roughly, a model for a subject domain rather than a specific application. For instance, all large companies have employees, executive boards, finance and human resources departments. We can create a model about it once and then sell it to companies that don't have enough in-house modelling skills to do it themselves. The enterprise models and ontologies have in common that they try to save you modelling time though re-use of existing part-solutions that other people have figured out before you. We'll get back to this in the next chapter.

The bottom-up approaches are variations of the first category of determinants to answering the question 'how do you model?', and of those, the data or sample instance approach in particular. They may be either a few manually constructed examples, like the book authors examples in Fig. 4.1, or taken from data sets that exist in another format, such as a spreadsheet. I used the sample instance approach with two domain experts—alternatively called 'clients'—at the time they didn't understand the diagrammatic notation and did not like reading text, quite unlike a third domain expert who brushed aside my carefully laid-out pretty diagrams in favour of the pseudo-natural language text generated from the diagram. Those two domain experts understood their data well, but knew insufficient about the conceptual modelling with its notation at the beginning of the project. They were in the life sciences and familiar with biological diagrams. A modeller, however, needs to get the domain expert to understand what's going on to validate the model in one way or another. To get them to understand the model, I tried to recall how I was thinking about data in those days when I conducted experiments in the microbiology lab. We worked with data and tried to find patterns. It doesn't have to be different for conceptual modelling. So, I took them by the hand from what I assumed they knew to what I wanted them to learn. Give them sample data, show the pattern, and map that to the notation of said pattern. In a way, it's the same way ER is normally taught, starting from natural language analysis just because some people assumed that's what students know best. That may not be the case in general, and particularly so in the sciences where most students don't like to read anyhow, and even more

4.2 How to Develop a Conceptual Data Model

Table 4.1 Mock data for Kathy's clothes emporium: valid and invalid

Valid	
Employee	Team
Joseph	D-rocks
Jane	D-rocks
Julia	D-rocks
Jane	Addmi
Josh	Addmi
Jane	Sales2
John	Sales2

Invalid	
Employee	Team
Joseph	D-rocks
Jane	D-rocks
Jane	Addmi
Jane	Sales1
Jane	Design
Julia	–

so where the medium of instruction isn't in the student's first language. A research problem I'll leave to computer science education or psychology or cognitive science researchers to solve.

My scientist experts knew their data better than they did natural language, or they preferred it over natural language. For instance, in the example in the previous section about developing an EER diagram for Kathy's clothes emporium, there's "have to work in teams of more than one employee" early on and at the end "can be a member of at most 3 teams" that had to be pieced together. Let's say you did not express that in natural language, be it written or spoken, nor did the domain expert, but instead you'd show them three tables, asking which of the options resembles the pattern of their data and how it violates the other(s). In other words: what is permitted in the data and what not. An example is shown in Table 4.1:

The valid data adheres to the constraints of the example. The invalid sample data violates the constraint of 'more than one' team member, since Addmi and Sales1 only have one member in the invalid sample data on the right. It also violates the 'at most 3' constraint because Jane is member of 4 teams. And it violates the mandatory constraint of team membership, as Julia isn't a member of any team yet. Or: with sample data we can figure out to a certain extent the constraints that have to be included in the model. In most cases, the constraints are simpler with just the data patterns for 1:1, 1:n, 0:n, and n:m cardinality constraints and you could prepare the valid and invalid tables accordingly. Alternatively, there may be an iteration: the client shows some data, the modeller observes the constraints, tweaks it and asks 'is this (dis)allowed?', and the yes/no answer finalises the constraints.

This example- or instance-based approach can be, and already has been pushed further. Instead of an on-the-side activity for validation of the model, it can be made into a part of the methodology, as with the Fully Communication-Oriented Information Modeling (FCO-IM) flavour of ORM.[15] That envelope can be pushed, with, among others, Test-Driven Development. This idea originates in programming, where the basic idea is that first a test is specified that must pass if all is well. It is expected to fail initially, then you add or modify a piece of code, and then the test passes. For conceptual model design, this may take the form of specifying statements about actual or hypothetical instances and checking whether the model allows them to exist. For instance, with Bicycle(mariasbicyle1), for it to pass, I need to have a class Bicycle in my model as a minimum. Instead of illustrative sample data to assist with determining what should be included in the model, the instances have become a model's scope specification and a quality control mechanism for it.

These bottom-up approaches are manual efforts. There are two main ways to automate the procedure: reverse engineer the model from relational data that you may have, or mine the salient entities from a suitable text corpus. The former requires well-structured data already, which is what we wanted to achieve with the modelling, and so thus don't have it yet—if we did, we didn't need to do all this modelling. That reasoning holds within the traditional scope of sequential design from model to database implementation, but there are re-inventions for using conceptual data models to help with formulating queries over the database, create dynamic documentation to match an adaptive database structure that optimises according to the kind of things most users want to get out of it, or optimise executing the query.[16] The reverse engineering task is well-known, was popular in the late 1990s, and meanwhile considered to be solved. It uses a combination of the database structure and that pattern-finding, not based on mock data but on trawling the data in the database and inferring minimal constraints from them. The second automation option has the start-up costs of, first, creating that corpus for your specific domain of interest, and then good natural language processing and summarisation algorithms to filter out what the key entities and relations are. Neither are cheap solutions, nor do they describe the entire process, especially, not that part of whether the conceptual data model's content is complete and correct.[17]

As final remarks, it can be argued that further research is needed on such bottom-up and top-down approaches to improve on the outcomes, but it's hard to do. From a methodological viewpoint, to show that a new method is better, the main problems to overcome are finding enough willing subjects that are a fair and fair-

---

[15] Described in (Bakema et al. 2005).

[16] Examples include ontology-based data access, of which a gentle introduction and application about food in the Roman empire is described by Calvanese et al. (2016). The first results on dynamic updates are reported by Zäschke et al. (2013).

[17] The sample data approach is part of the ORM modelling processes and tools. A key reference on the idea of test-driven development in programming is written by Beck (2003) and the general idea was ported to conceptual data models for UML by Tort et al. (2011) as well as for the models we'll see in the next chapter as first proposed in (Keet and Lawrynowicz 2016) and extended in (Davies et al. 2017).

4.2 How to Develop a Conceptual Data Model

sized sampling of the modelling population—which therewith excludes the easily accessible computer science students—and an agreeable way to determine what exactly a 'good' conceptual data model is. So even if you think you've come up with a novel terrific way for conceptual data model development, it will be hard to convince colleagues with irrefutable evidence that your methodology is better than the existing ways. But exactly that is a prerequisite for science to be advancing, to demonstrate that the novel proposal improves the state of the art—and for academics getting a paper accepted as a result of it. (No evaluation, no paper, no brownie points to make a career in academia.) Perhaps there is no very convincing way to advance conceptual data model design methodologies. On the bright side, however, there are several tried and tested guidelines, which count for something already and those guidelines most certainly are well ahead of the absence of even one tested methodology to develop mind maps and biological models.

4.2.2.1 Dance and Conceptual Data Models

Having traversed the main conceptual data modelling language families and ways to develop one, from high-level generic guidelines to specific ones for the model construction, how could this work for out for our running domain of dance? We could choose a topic covering a part of the mind map of Chap. 2 or do something with the model for the dancing lyrebird of Chap. 3. A model for the former could be to imagine a dance school whose secretary wants a database about its dancers, like which class they are enrolled in, at what level, and in which contests couples took part, if any, and whether they rented streaming video for online classes. A model for the latter would have as aim to store the scientific data collected about the lyrebirds and to afford easy querying functionality to analyse the data. Based on the research methodology described by Dalziell and co-authors, such a database stores data about identifiable lyrebirds, whole video and audio clips at a location at a particular date, annotation terminology, and annotated video and audio fragments.

The first observation for both scenarios is that there is no neat text paragraph with a description for a database or application available. This leaves any of the top-down or bottom-up approaches. For a subject domain as specialised as a lyrebird dance database there is doubtlessly no readily available enterprise model to base it off, so a top-down approach is out of the question. Generic models for schools will be available, but they're too costly for the town's dance school, and so we're left with a variant of bottom-up design approach: get to know the domain, collect sample data, and probe the client to figure out what they need. The sample data will assist ticking off step 1 of the conceptual schema design procedure.

Concretely, then, the model design. Since a database for the prospective dance school is likely to be more widely useful, I leave that for a free sample online and continue here with the multimedia database for the scientific data, based off Dalziell et al.'s data description in the article. The first step is to convert that into elementary fact types, of which a selection is included in Table 4.2.

For step 2, drawing the fact types and adding identifiers for the entity types, we do need to make up our mind which flavour of conceptual model we're going to draw. To show that procedure can just as well be applied to ER diagrams, we'll go

Table 4.2 Illustrative elementary fact types for the lyrebird multimedia database

multimedia file #1 records toby	multimedia file #1 was recorded on 7-7-2017
multimedia file #2 records cory	multimedia file #2 was recorded on 7-7-2017
multimedia file #3 records toby	multimedia file #3 was recorded on 10-7-2018
bird1 has as pet name toby	multimedia file #1 has part fragment 1.1
bird2 has as pet name cory	multimedia file #1 has part fragment 1.2
.....	etc.

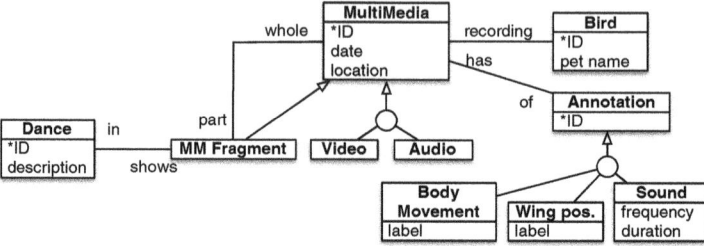

Fig. 4.9 The ER diagram 'in progress' after step 2 of the conceptual schema design procedure

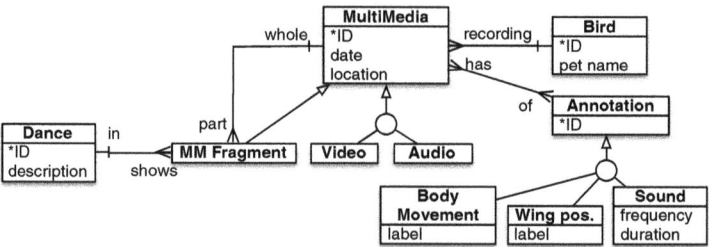

Fig. 4.10 The ER diagram 'in progress' after step 4 of the conceptual schema design procedure

with that. A first pass of just the entity types, relationships, and attributes is shown in Fig. 4.9.

There is no duplication, so step 3 can be checked as complete as well. Step 4 concerns the cardinality constraints on the relationships. There may be multiple recordings for each bird and there is only one bird for each recording; each recording has multiple annotations that can be reused across recordings; and each recording can be split up into multiple fragments. This gets us to the second partial diagram, shown in Fig. 4.10.

Now, in step 5, we need to add the mandatory constraints. Participation is mandatory for the birds and multimedia files, but not all multimedia files need to have fragments nor annotations. And there we obtain the next version, as shown in Fig. 4.11.

Step 6 asks for the first set of advanced constraints: anything on attribute values and subtyping. There are no immediately obvious upfront constraints on the attribute

4.3 Limitations

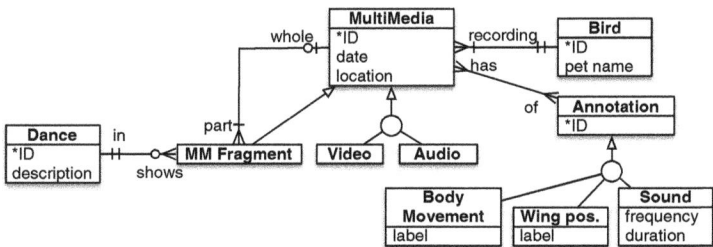

Fig. 4.11 The ER diagram 'in progress' after step 5 of the conceptual schema design procedure

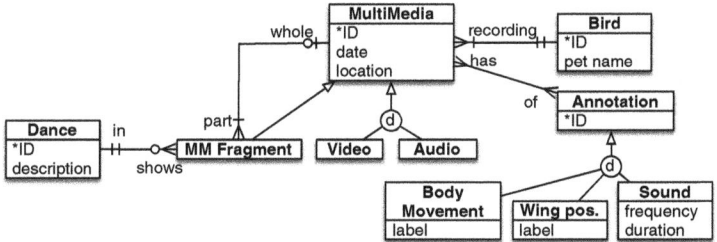

Fig. 4.12 The ER diagram after step 6 of the conceptual schema design procedure

values, so we leave that be. We do have subtypes, and some of them are surely disjoint, such as the audio and video, so we add that. It brings us to Fig. 4.12.

ER does not have ORM's ring constraints, so step 7 may not sound applicable to ER. But also: ER has features that don't seem to be covered by ORM's procedure. Specifically, ER has weak entity types, which was relegated to a footnote earlier in this chapter. Yet, ORM is strictly more expressive than ER, so it can't have been missed, either. What's going on? Merely a terminological difference, it turns out. ER's weak entity type is equivalent to a special external uniqueness constraint in ORM, which is part of Step 4. We don't need it in this scenario. And so this completes the procedure and Fig. 4.12 is our final model.

4.3 Limitations

Since conceptual data modelling and their languages solve problems of those biological models, would it nonetheless be possible to complain about them as well, aside from bickering among the flavours? Of course; we just have to move the goalposts a little. There are three main aspects about which one can argue that the conceptual data modelling languages fall short.

First, they're imprecise on their own as diagramming languages. The UML specification as standardised by the OMG organisation is mostly text and diagrams for over 700 pages, with phrases that are grating to computer scientist for their imprecision. For instance, a language feature can be a so-called "semantic variation point", or: you can choose what you want it to mean! UML's aggregation association is a semantic variation point. Also, they all have relationship components and relationships, and ORM also has fact type readings that are neither, but which elements have to be named and which of them are implemented in the database or object-oriented code is up to the tool developer. Also, it's not prohibited to create a model where a supertype has more attributes than its subtype, which violates the principle of subsumption, but a plain conceptual data modelling tool will not flag it as incorrect. Or to assert a subtype to have an identifier that is different from its supertype, which is also wrong as the subtype must inherit the supertype's identifier. Yet, for anything that the modellers or domains expert can't figure out, they kick the can down the road to the software developers to decide what to do with the imprecision. If the developer chooses something the modellers or domain experts didn't want after all, it will be costlier to re-engineer the software than if the analysts would just have made up their mind in the early stages of the software development process.

There's a solution to that problem: devise a formal, logic-based underpinning so as to enforce the precision in the language itself and reason over the model automatically to derive the implicit constraints. The precision was already part of ORM with Terry Halpin's thesis that he completed in 1989. He used first order predicate logic for that, which is not convenient for computation. The research direction inclusive of finding an algorithmic way to check correctness started nearly 30 years ago with Maurizio Lenzerini at the La Sapienza University in Rome, Italy. He and his students—meanwhile accomplished professors themselves—made early major advances with formalising fragments of ER and later also UML Class Diagrams once that became popular and created a hybrid notation in the ICom tool around the turn of the millennium. The theory and techniques were well ahead of the times. It spawned two further efforts: ORM with the ORMiE plugin that extends aforementioned NORMA plugin, and crowd2.0 that supports the core fragment of all three language families.[18] Put differently: it can be fixed. Still, none of that is mainstream in conceptual modelling. It's easier to 'sell' that work elsewhere, as we shall see in the next chapter. I have suspicions as to why the conceptual modellers don't want to avail of all those advances, but the reasons for refusal of its uptake have not been investigated, and so it remains guesswork and hoping that the modellers will see the light.

Second, there are model quality beyond making sure there are no conflicting entity types and the related issue of reinventing the wheel. The latter issue has been recognised and that's where the top-down approaches come in, which can address it partially or wholly with enterprise models and modelling solutions of

---

[18] For the first major results, see Calvanese et al. (1998), who formalised part of ER in the Description Logic $\mathcal{DLR}$, extended it for UML Class Diagrams (Berardi et al. 2005), and then also for many more conceptual data modelling language fragments and logics; see (Fillottrani and Keet 2021) for additional references. Icom was presented first in (Franconi and Ng 2000); ORMiE (Sportelli and Franconi 2019) and crowd2.0 (Braun et al. 2021) are recent developments.

the various snippets. They are generic patterns that you can take and tweak for your own scenario. However, this works only for common domains. If the domain isn't common or the enterprise models are out of reach, you or the modellers you hire will have to do the heavy lifting of modelling from scratch. This leaves a gap to be filled, to offer guidance on what to do then. Run-of-the-mill conceptual data model per sé doesn't have a solution, but snippets are, in fact, available. We'll need to get those from elsewhere, beyond conceptual data modelling. We'll see that in the upcoming chapters.

Third, we're running into problems with data integration and match-making of content across databases. It's one of those intractable problems that started in earnest in the mid 1990s, and it's still not easy to do. Instead of going with your hands in the mud poring over database tables and program code to figure out how to combine the data, it's possible to do that at the level of the conceptual data models. At that level, a modeller can specify what is the same, what's different, what are the subtypes, how to convert and so on, possibly informed by an algorithm that first tried to find candidate matches. That works for two models, which requires one set of such mappings. If you have three models, all with 1-to-1 mappings, there are 3 sets of mappings, with four models, there are 6, and so on. That's easily getting too hard to oversee and maintain, and according to Michael Stonebraker of the Massachusetts Institute of Technology and Turing award winner for his contributions in database management technologies, and Ihad Ilyas at the University of Waterloo, and their experiences with the Tamr company, integration projects nowadays have to integrate a number of data sources that is in the double digits.[19] One-to-one model mappings aren't a solution then anymore; they've become a problem. We need other solutions for that. That's not a fault of conceptual data models; the new demands have outgrown the original aims. Let's get on with that in the next chapter.

References

Artale A, Parent C, Spaccapietra S (2007) Evolving objects in temporal information systems. Ann Math Artif Intell 50(1–2):5–38

Bakema G, Zwart JP, van der Lek H (2005) Volledig Communicatiegeoriënteerde Informatiemodellering FCO-IM. Academic Service

Beck K (2003) Test-driven development: by example. Addison-Wesley Professional

Berardi D, Calvanese D, De Giacomo G (2005) Reasoning on UML class diagrams. Artif Intell 168(1–2):70–118

Braun G, Fillottrani PR, Keet CM (2023) A framework for interoperability between models with hybrid tools. J Intell Inform Syst 60:437–462

Braun GA, Marinelli G, Gavagnin ER, Cecchi LA, Fillottrani PR (2021) Web interoperability for ontology development and support with crowd 2.0. In: Zhou Z (ed) Proceedings of the Thirtieth International Joint Conference on Artificial Intelligence, IJCAI 2021, ijcai.org, pp 4980–4983

---

[19] (Stonebraker and Ilyas 2018).

Calvanese D, De Giacomo G, Lenzerini M (1998) On the decidability of query containment under constraints. In: Proceedings of the 17th ACM SIGACT SIGMOD SIGART Symposium on Principles of Database Systems (PODS'98). ACM, New York, pp 149–158

Calvanese D, Liuzzo P, Mosca A, Remesal J, Rezk M, Rull G (2016) Ontology-based data integration in epnet: production and distribution of food during the roman empire. Eng Appl Artif Intell 51:212–229

Codd EF (1970) A relational model of data for large shared data banks. Commun ACM 13(6):377–387

Davies K, Keet CM, Lawrynowicz A (2017) TDDonto2: A test-driven development plugin for arbitrary TBox and ABox axioms. In: Blomqvist E, Hose K, Paulheim H, Lawrynowicz A, Ciravegna F, Hartig O (eds) The Semantic Web: ESWC 2017 Satellite Events. LNCS, vol 10577. Springer, Berlin, pp 120–125

Fillottrani P, Keet CM (2021) Evidence-based lean conceptual data modelling languages. J Comput Sci Technol 21(2):e10

Fillottrani PR, Keet CM (2014) Conceptual model interoperability: a metamodel-driven approach. In: Bikakis A et al (eds) Proceedings of the 8th International Web Rule Symposium (RuleML'14). LNCS, vol 8620. Springer, Berlin, pp 52–66

Fowler M, Parsons R (2010) Domain-specific languages. Addison-Wesley Professional

Franconi E, Ng G (2000) The i.com tool for intelligent conceptual modeling. In: Bouzeghoub M, Klusch M, Nutt W, Sattler U (eds) Proceedings of the 7th International Workshop on Knowledge Representation meets Databases (KRDB 2000), CEUR-WS, vol 29, pp 45–53, Berlin

Halpin T (1989) A logical analysis of information systems: static aspects of the data-oriented perspective. PhD thesis, University of Queensland, Australia

Halpin T (2001) Information modeling and relational databases. Morgan Kaufmann Publishers, San Francisco

Halpin T, Morgan T (2008) Information modeling and relational databases, 2nd edn. Morgan Kaufmann

Keet CM (2017) Natural language template selection for temporal constraints. In: CREOL: Contextual Representation of Events and Objects in Language, Joint Ontology Workshops 2017, CEUR-WS, vol 2050, p 12

Keet CM, Artale A (2010) A basic characterization of relation migration. In: Meersman R et al (eds) OTM Workshops, 6th International Workshop on Fact-Oriented Modeling (ORM'10). LNCS, vol 6428. Springer, Berlin, pp 484–493

Keet CM, Berman S (2017) Determining the preferred representation of temporal constraints in conceptual models. In: Mayr H et al (eds) 36th International Conference on Conceptual Modeling (ER'17). LNCS, vol 10650. Springer, Berlin, pp 437–450

Keet CM, Fillottrani PR (2015a) An analysis and characterisation of publicly available conceptual models. In: Johannesson P, Lee ML, Liddle S, Opdahl AL, Pastor López O (eds) Proceedings of the 34th International Conference on Conceptual Modeling (ER'15). LNCS, vol 9381. Springer, Berlin, pp 585–593

Keet CM, Fillottrani PR (2015b) An ontology-driven unifying metamodel of UML Class Diagrams, EER, and ORM2. Data Knowl Eng 98:30–53

Keet CM, Lawrynowicz A (2016) Test-driven development of ontologies. In: Sack H et al (eds) Proceedings of the 13th Extended Semantic Web Conference (ESWC'16). LNCS, vol 9678. Springer, Berlin, pp 642–657

Keet CM, Ongoma EAN (2015) Temporal attributes: their status and subsumption. In: Köhler H, Saeki M (eds) Asia-Pacific Conference on Conceptual Modelling (APCCM'15), Conferences in Research and Practice in Information Technology, CRPIT, vol 165, pp 61–70

Ongoma EAN (2015) Formalising temporal attributes in temporal conceptual data models. Msc thesis, Department of Computer Science, University of Cape Town, South Africa

Parent C, Spaccapietra S, Zimányi E (2006) Conceptual modeling for traditional and spatio-temporal applications—the MADS approach. Springer, Berlin

Reinhartz-Berger I, Sturm A, Clark T, Cohen S, Bettin J (2013) Domain engineering: product lines, languages, and conceptual models. Springer, Berlin

Shoval P, Shiran S (1997) Entity-relationship and object-oriented data modeling—an experimental comparison of design quality. Data Knowl Eng 21:297–315

Shunmugam T (2016) Adoption of a visual model for temporal database representation. M. IT thesis, Department of Computer Science, University of Cape Town, South Africa

Sportelli F, Franconi E (2019) A formalisation and a computational characterisation of ORM derivation rules. In: Panetto H, Debruyne C, Hepp M, Lewis D, Ardagna CA, Meersman R (eds) On the Move to Meaningful Internet Systems: OTM 2019 Conferences—Confederated International Conferences: CoopIS, ODBASE, C&TC 2019. LNCS, vol 11877. Springer, Berlin, pp 678–694

Stonebraker M, Ilyas IF (2018) Data integration: the current status and the way forward. IEEE Data Eng 41(2):3–9

Tort A, Olivé A, Sancho MR (2011) An approach to test-driven development of conceptual schemas. Data Knowl Eng 70:1088–1111

Zäschke T, Leone S, Gmünder T, Norrie MC (2013) Optimising conceptual data models through profiling in object databases. In: Ng W, Storey VC, Trujillo J (eds) Conceptual Modeling—32nd International Conference, ER 2013, Hong-Kong, China, November 11–13, 2013. Proceedings, Springer, LNCS, vol 8217, pp 284–297

Ontologies and Similar Artefacts 5

The perfect is the enemy of the good.

— *Voltaire*

As you may expect by now, the plan for this chapter is to solve the actual and newly created limitations of the types of models introduced in the preceding chapters. We'll get to that. First, as a "told you so" upfront: we are about to enter a field that is solidly within the realm of 'back office' or 'back-end solutions' of IT systems. Normally, no end user of any system would be informed about such models to exist and that at least one of them is used in a system. Yet, *ontologies* most certainly are used. An ontology powers the 'intelligent' part of an intelligent information system and it's used in various AI systems; IBM's Watson, of the *Jeopardy!* game fame, uses multiple ontologies, even. It may be in order to resuscitate an early 2000s publicity chuckle we had about ontologies: the confidence-inspiring Intel logo stickers with "Intel inside" stuck on PCs had been modified and stated rather "ontology inside". In the early years of PCs, most people had no clue what it really meant that the core of the central processing unit (CPU) was made by Intel, what exactly it was doing inside the computer, where it was located, or how the CPU was doing its things. But we all knew that a computer can't do without one. We're going to break open that CPU and look inside, as it were, but then for ontologies. Where the analogy with a CPU breaks down, is that most applications can do without ontologies—just not as well as with them. From searching the Web to optimising data mining experiments to putting a database 'on steroids' for better query management and question answering systems: the system with an ontology inside always wins from the one without an ontology inside.

I stumbled onto ontologies in 2003 during that same honours project that made me dust off ORM. I didn't want to model a complicated piece of information from scratch and was hoping someone had solved the problem before me and I'd be able to reuse it. Searching online, I came across something that had the information in part. That something was an ontology, the Gene Ontology, to be precise. More generally, ontologies can solve issues of conceptual data modelling from the previous chapter. They enjoy the rigour that was lacking with conceptual data models, there are additional quality assurances with science-based methods, and their original aim was to solve that data integration problem, which they indeed do contribute to. Not just that, other fields within computer science, software engineering, and IT took note and adopted ontologies as models for their own purposes.

So, what is that purported wonder potion called 'ontologies'? Very informally, an ontology is an artefact that contains the entity types, the relationships among them, and the constraints that hold over them for a specific subject domain. It almost sounds like a description of EER diagrams, but it is not. There are two key differences. One tacit difference is that knowledge is represented formally, in a suitable logic, and normally in such a way that the computer can process it. This imposes a few limitations and offers new opportunities that we'll get to in due course. The other difference looks harmless yet has far-reaching consequences: the 'for a specific subject domain' in that informal sentence. This entails that whatever is represented in the model will have to hold *across multiple* applications in the same domain, which is fundamentally different from the conceptual data models that are intended only for *one particular* application.

To start lifting the tip of the veil of the latter difference, consider a scenario of wanting to develop a database for the local rugby club that you may be the secretary of, so as to better manage the memberships and the equipment, and another scenario where the MyPlayers organisation for professional rugby players requires a model that will work for *all* rugby clubs in South Africa or the Rugby Union internationally for all clubs in all member countries, or even to capture knowledge about sports clubs generally. For your local club, you make decisions for data storage that are relevant to that application, which may not hold for another rugby club, let alone a club for another sport. For instance, whether, and if so, how, to store each member's address and your club may have a constraint that each member is permitted to be member of at most one team. Another rugby club may be more lenient; like allowing a player to be member of a team that plays competitively regularly and, given their age, that they're also a member of the 'oldies' team. Such details are specific to the application for each club. Compare that to general knowledge about sports clubs: they have members and they have teams, members play in teams, there are teams that play in matches in regional, national, and international competitions, and so on. It is such general knowledge that goes into an ontology. Similarly, an ontology about dwellings would have general knowledge about the various types of dwellings, that a house has a roof as part and doors as well, and is made of a particular type of material, among other things. A database for a specific housing project goes into the specifics of only those houses they're going to build, such as a row of townhouses having exactly two exterior doors, one bathroom, and a corrugated iron roof among other parts.

We shall flesh this out in the remainder of this chapter, so that you'll get a better idea of what an ontology is, the sort of things that are represented in such type of models, what you can do with it, three success stories, and a few notes on how to create one yourself. There are also shortcomings, as usual—nothing is perfect. And, just like for the models we've seen in the previous chapters, there are whole books devoted to just these types of models, whereas here we have to make do with just one chapter. Especially for this chapter that is harder to achieve for me, since I wrote a textbook about ontologies that's aimed at a readership of honours (4th year) and postgraduate students in computer science.[1] It's easy to get carried away with the subtopics and yet the readership and aims are dissimilar. What both have in common is that we'll first lift the tip of the veil of what ontologies are.

5.1 What Is an Ontology, the Artefact?

It's a terrible question to answer if you're looking for a one-liner of higher precision than the aforementioned description in the introduction of this chapter. The main reason why it is hard to get a water-tight, yet simple, description is that ontologies are used across diverse disciplines, and they all have their own assumptions to go with different terms that could potentially be used in such a definition. There are people from philosophy, cognitive science, psychology, and linguistics from the humanities side of the spectrum, and logicians, computer scientists, and software developers from science and engineering, and then a bunch of domain experts from any field who like to work with ontologies as well. A discussion is included in my textbook;[2] here, we'll stick with the informal description as a sufficient approximation, and we'll colour in the details as we progress.

Given that ontologies are a notch up from conceptual data models, we'll need to increase the technical aspects a notch compared to the previous chapter. If you prefer to read success stories before such a commitment, this may be feasible for at least the first one in Sect. 5.2.1, which starts humbly in the late 1990s and is going ever stronger since. But slogging through this section will be rewarding: there's a dash of logic in store, a pinch of reasoning, and a wisp of analytic philosophy to complete the section.

5.1.1 Syntax and Semantics

Let's start with a small example and work our way up from there. Figure 5.1 presents two diagrammatic representations of an ontology about houses, first in UML-like notation and then in an ad hoc graphical notation. There's substantial abuse of notation, stemming from the motivation to not invent yet another notation when we can reuse what a segment of the user base is already familiar with. One reason

---

[1] The book was published in July 2018, with a v1.5 in February 2020 (Keet 2018). It's an open textbook, i.e., accessible for free as a pdf, from https://people.cs.uct.ac.za/~mkeet/OEbook/, and hard copies can be bought from online retailers, as it is also published by College Publications.
[2] (Keet 2018).

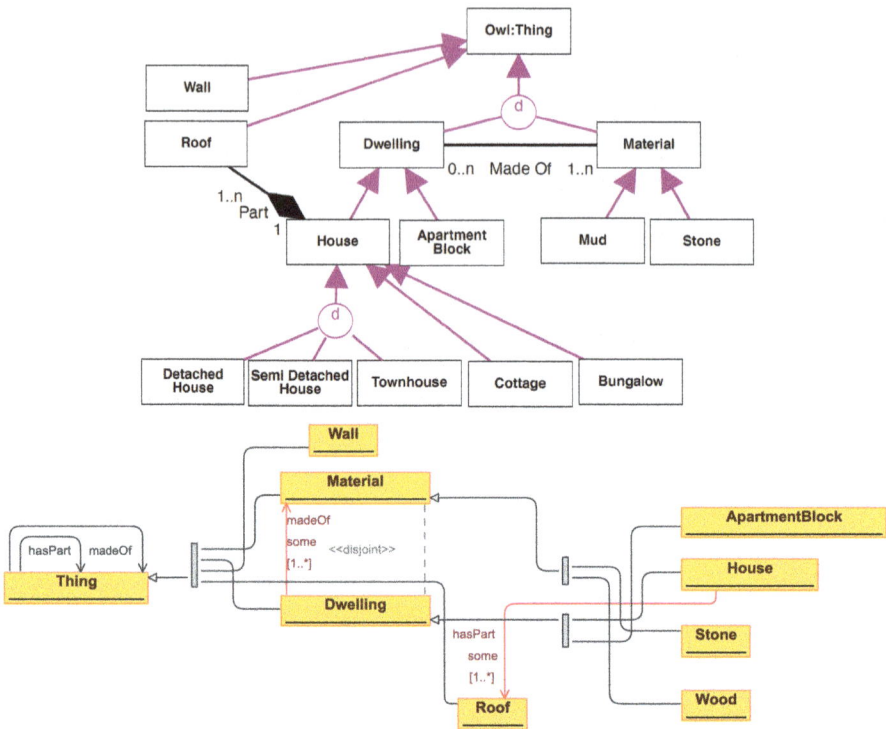

Fig. 5.1 Example of a very small toy ontology about houses: the diagram on the top-half was created with the `crowd2` online editor and the diagram at the bottom-half was automatically generated by the OWLGreD tool by loading that serialised version of the ontology (minus the subclasses of House). See Fig. 5.2 for a selection of the formalisation and serialisation

to not reuse the visual notation of conceptual data modelling languages is to avoid confusion that they might be all the same and another reason is that their language features aren't the same.[3] Notation abuse has largely won the contest. A new type of model on the block solves its communication task by relying on a solution that was proposed for another aim and whose problems it was meant to solve. Let's disentangle that as we progress.

The first of the two key differences is what's behind those diagrams, of which a section is shown in Fig. 5.2: a logic-based reconstruction of part of Fig. 5.1. Each

---

[3] To be precise on the mismatch, features that conceptual data modelling languages have but popular ontology languages such as OWL do not: n-ary relations where n may also be larger than two, roles that objects play in a relationship, and multi-attribute identifiers or external identifiers. Conversely, what OWL has but conceptual data modelling languages typically do not: defining entity types and disjointness between classes (without a parent class). An example of a recent 2D graphical notation for ontologies is described in (Lembo et al. 2022) and there are also proof-of-concept tools with 3D visualisation techniques (Dooley and Hsiao 2019).

5.1 What Is an Ontology, the Artefact?   85

Part of the ontology in DL notation

Dwelling ⊑ ∃madeOf.Material
Dwelling ⊓ Material ⊑ ⊥
House ⊑ Dwelling
House ⊑ ∃hasPart.Roof
SemiDetachedHouse ⊑ House

Part of the ontology serialised in OWL functional style syntax

```
# Class: :Dwelling (:Dwelling)
SubClassOf(:Dwelling ObjectSomeValuesFrom(:madeOf :Material))
DisjointClasses(:Dwelling :Material)
# Class: :House (:House)
SubClassOf(:House :Dwelling)
SubClassOf(:House ObjectSomeValuesFrom(:hasPart :Roof))
# Class: :SemiDetachedHouse (:SemiDetachedHouse)
SubClassOf(:SemiDetachedHouse :House)
```

Fig. 5.2 Example of a very small ontology about houses: a fragment of the logic on the left and the same knowledge serialized such that the computer can process it in applications on the right. See Fig. 5.1 for diagrammatic renderings

Table 5.1 Sampling of correspondences between a common syntax notation in first-order predicate logic and Description Logics and below that, one or more informal wordings of the meaning and terminology used to describe it

First-order predicate logic (FOL)	Description logics (DL)
$\forall x(C(x) \rightarrow D(x))$	$C \sqsubseteq D$
Each instance of C is also an instance of D; C is a subclass of D	
$\forall x(C(x) \rightarrow \neg D(x))$	$C \sqsubseteq \neg D$
No instance of C is also an instance of D; Cs are not Ds; C and D are disjoint	
$\forall x(C(x) \land D(x))$	$C \sqcap D$
All those instances that are an instance of both C and D; the intersection of C and D	
$\forall x(C(x) \rightarrow \exists y(R(x,y) \land D(y)))$	$C \sqsubseteq \exists R.D$
Each instance of C relates through R to some instance of D; Each C R at least one D	

element in the graphical rendering maps onto an element or an axiom in the logic, and vice versa for this diagram. That "Each House has as part at least one roof", with the two rectangles, the line, the black diamond, and the 1..* in Fig. 5.1 can be formalised in Description Logics, a popular family of logics for ontologies, as

$$\text{House} \sqsubseteq \exists \text{hasPart.Roof} \qquad (5.1)$$

and in first order predicate logic syntax as

$$\forall x(\text{House}(x) \rightarrow \exists y(\text{hasPart}(x,y) \land \text{Roof}(y))) \qquad (5.1')$$

Different notations, same meaning. The "Each" in the controlled natural language sentence is governed by the "∀" symbol, the "at least one" by the "∃", and the "has as part" by the hasPart. A couple of those syntax mappings are listed in Table 5.1.[4] The different notations for the logics sound like it may be ripe for a turf war alike on conceptual data modelling. There hasn't been one. Mutterings about preferences

---

[4] See the Description Logic handbook (Baader et al. 2008) for mappings between the two, and much more. A comprehensive introduction to first order predicate logic can be found in, e.g., Hedman (2004).

for logics can be heard, but also here it has settled into a live and let live and any conversions happen behind the scenes if needed.

One advantage of the logic is the precision. That "semantic variation point" in UML? Here it must be specified. Take, for instance, UML's aggregation association, which is drawn merely with a black diamond shape. However, by its assumed behaviour of object destruction where the parts are deleted when their whole is deleted and a class can play both the part and whole role at the same time, it must be transitive. That can also work the other way around: a part of a part is also a part of the whole, like that the brick that is part of the wall that is part of the house, is also part of the house. In Description Logic notation, either the Trans(part) notation can be used to capture that or a role chain to link at least three objects in a chain:

$$\text{part} \circ \text{part} \sqsubseteq \text{part} \tag{5.2}$$

The logic-based foundation can be a disadvantage as well: now you must make up your mind about the details.

That precision we only really obtain because the logic and its corresponding plain text notation for computation are not just syntax, as if it were merely a notation that I'd be making up to chase away people who don't like mathematics and who would rather draw diagrams. There's a formal semantics to it as well. That is: those 'squiggles', like the "∃" and "∧" in Eqs. (5.1), (5.1'), and (5.2), mean something. There are several ways to give meaning to it; for the logics we use for ontologies, it is most often a so-called *model-theoretic semantics*—with semantics to be understood within the realm of logics, not natural language or meaning in a subject domain. Dissecting the term, there's something that's called 'theory', which would be our ontology, and something that's called a 'model' that has the (possible or actual or both) instances that are supposed to adhere to what we declared in the theory. Indeed, confusingly, the word 'model' in logics is not what it refers to everywhere else in conceptual modelling when people talk about models and modelling. The 'model' in logics may be, e.g., the data stored in database tables that adhere to the type-level specification of the database schema. I'm going to tweak the logicians' terminology in one aspect, for the sake of readability and coherence across the book's content, and call such a 'model' a particular *structure* of the theory that is our ontology.[5]

That cleared up, if a type-level specification (i.e., theory, ontology, the notion of model otherwise used throughout the book) states that each house has exactly one geyser, then any particular structure, as stored in a database of a housing estate developer, say, cannot have ⟨house1, geyser1⟩ and ⟨house1, geyser2⟩ in the same table, under the assumption that geyser1 and geyser2 are two

---

[5] It is acknowledged that this is not very precise, though it is true that models in that sense are mathematical structures. Section 5 in the Stanford Encyclopedia of Philosophy entry on model theory by Hodges (2022) tries to trace the origins of the different usage—and confusion it can cause—of the term 'model'.

5.1 What Is an Ontology, the Artefact?

distinct things, since 'two' violates 'exactly one'. If it were to do so, the whole package together—theory and structure—would be inconsistent. Phrased from another viewpoint: that particular structure must be a structure of another theory then, but surely not the one declared and tested against. That checking of whether a structure may exist takes into account the instances declared, and for the rest of them, it will generate temporarily instances for those entities that need to have one if it were to be instantiated, for the sake of checking whether they could exist. The theory tells us something about the 'world', or permissible worlds, we can construct with it, abstractly.

The definitions for the model-theoretic semantics are built up in multiple steps, which are long and tedious. I'll simplify that story to one with the more familiar notion of sets to sketch the gist of it. We can declare House to refer to all individual houses. In first order logic notation that is House(x) where the x is a variable, standing in for any house. We can do likewise for other classes of objects, such as Mud. An element we have seen in axiom (5.1') is the "∧", or "⊓" in Description Logics notation, which is the 'and', and is also called the intersection or the overlap: those individuals that are both whatever is declared left from it and whatever is declared on the right of it. For instance, House ⊓ Mud denote those objects that are both in the set of houses and in the set of instances of mud. We could name the set of those objects, say, MudHouse. Then it gets interesting. How should we relate House and Mud? We could state that a mud house is both a house and mud, like

$$\text{MudHouse} \sqsubseteq \text{House} \sqcap \text{Mud} \qquad (5.3a)$$

or, rather, add to our logical theory (the ontology) that a mud house is a house that is made of mud:

$$\text{MudHouse} \sqsubseteq \text{House} \sqcap \exists \text{madeOf.Mud} \qquad (5.3b)$$

The logic doesn't care about which one of the two you choose; logic has no opinion. The consequence of the choice we make, however, is going to be different for each axiom. Mere syntax and semantics of the logic with the current small theory we have is not enough to determine which one is better. As spoiler alert: we'll see why (5.3b) is better shortly. A first hint is to add content to our theory, by declaring that House and Mud are disjoint, since a house is a 3-dimensional object that can be counted whereas mud is stuff and can be counted only in amounts of the stuff. We do that with the "¬" symbol that means 'not':

$$\text{House} \sqsubseteq \neg \text{Mud} \qquad (5.4)$$

That is, an object in one set can't also be an object in the other set, or: those two sets can never have a non-empty intersection, i.e., they are disjoint sets—variation in describing it, same meaning.

This link between theory and structure, on what instances we can have and how they relate, gives meaning to the notation with its elements and grammar. Only once the syntax is given meaning, we can say whether the statements (axioms) make sense. Of course, this happens only after checking that the syntax is correct. In a considerable number of projects, a modeller will be happy with this logic machinery as it is already. They shouldn't. Axiom (5.4) spells trouble for our theory so far, if it includes (5.3a). As a minimum, we need a way to find the troublesome parts of any theory. Luckily, logic can contribute more, especially within the realm of computational logic: *automated reasoning*. It can automatically find that axiom (5.4) and mud houses in the sense of (5.3a) don't agree, among other things. Let's look into that in the next section.

5.1.2 Automated Reasoning

Let me get the caveat out of the way upfront: we obtain that advantage of automated reasoning for ontologies subject to choosing the logic wisely. The logic has to have a *serialization* specified so that the computer can use the axioms in software applications, like depicted on the right-hand side in Fig. 5.2. For instance, instead of that "∃" that is called *existential quantification* and pronounced as 'there exists' or 'at least one' or 'some', that serialisation uses ObjectSomeValuesFrom in plain text. We can feed this plaintext format of the axioms into an application and do with it whatever we want, within the bounds of computation.

The serialisation is a start, but it's not enough for automated reasoning. It'll help us to go beyond the paper-based truth tables you may have come across in a mathematics course at university. That approach doesn't work for the logics and the sizes of the ontologies we work with. Ontologies easily can have thousands of classes and axioms, so it's a really dumb idea to create a table with all true/false permutations for each class and all the axioms it participates in, even when there's lots of computing power available. There are better strategies to obtain the outcome faster. Computer science outperforms mathematics on those strategies.

To live up to that claim, engineers designed and implemented automated reasoners. Such software tools take three kinds of ingredients: (1) the precise problem to solve, (2) a language that the computer can process, and (3) algorithms or procedures that do that reasoning for us. The problem is solving the task of answering the question "is this class satisfiable/can it ever have any instances?" with either a 'yes' or a 'no'. No sort-ofs, maybes, or to a degree, like you can't be roughly graduated either. The language is a logic; practically, the most widely used one for this task is a serialised Description Logic.[6] The algorithms have to do with the rules of inference for the language features, which we shall see shortly. That encoding of those rules is easy to do in a programming language like Python or Java, but the snag is in factors like which order to call them to apply it to the statements and how to do that computation efficiently.

---

[6] See the DL handbook (Baader et al. 2008) for the logics and the OWL specification for the serialisation (Motik et al. 2009).

To make a long story short and jump straight to the winner in surpassing truth tables from the paper-based mathematics classes: tableau algorithms. They're a way of deduction such that we do not have to enumerate all possible true/false combinations, but go in straight lines down in the tree to the sore spot(s), if there is one, or in the shortest possible way to demonstrating that a structure can be built. An example of a rule of deduction is 'if we have a conjunction, then each of the conjuncts must be true', which is then applied to the statements declared in the ontology. Let's consider axiom (5.3a) again, on each mud house being an instance of house and of (an amount of) mud. Each language feature, such as the "⊓" used in Eq. (5.3a), has a rule for how to interpret whatever we declared in our ontology into our structure instantiating our ontology. The automated reasoner applies all the rules to the ontology to try to build a structure consisting of possible instances, trying to find a way such that everything in the ontology can be instantiated without running into trouble with other parts of the ontology.

That may sound harmless, but that automated reasoner also finds axiom (5.4) in our theory, the one that states that house and mud are disjoint. What happens now? The reasoner will try to build a possible structure regardless. First, from (5.3a), it introduces a temporary entity trying to instantiate mud house, denoted with MudHouse(m1) to state that object m1 is an instance of the class MudHouse. That tableau rule for "⊓" is then applied, so then House(m1) and Mud(m1) must also hold. Axiom (5.4) states that if something is a house, as m1 is, it cannot be also an instance of mud. But we have Mud(m1)! This is a contradiction. Contradictions are bad. If we can't build a structure, then, by definition, we don't have a theory either, and, thus, in the strict sense, also no ontology. It's not only that m1 can't exist. It's the same story for any potential instance of MudHouse. There never can be any instance that is a mud house, according to this theory. More precisely: we say that MudHouse is *unsatisfiable*. The automated reasoner finds this for us. True enough that it's overkill for two small axioms, but consequences are not overseeable by humans once we get to many axioms, like 100 of them, or 5000, not to mention the thousands in a typical ontology.

More reasoning can be done. Among others, instance classification and related deductions. Let's tweak (5.3b) to define mud house, swapping "⊑" for "≡":

$$\text{MudHouse} \equiv \text{House} \sqcap \exists \text{madeOf.Mud} \tag{5.5}$$

Then if our logic theory has declared that Lindiwe's house is a house and m1 is still some mud, and the two relate as madeOf(Lindiwe's house,m1), the reasoner will deduce MudHouse(Lindiwe's house).

The one other key feature of automated reasoning is inferring the taxonomy, or deducing which class is necessarily a subclass of which other one(s). We modelled that manually in conceptual data models; here, we can get the computer to do that for us. Popular examples in the ontology literature to illustrate it are about pizzas, wine, or vegans and vegetarians. Among the examples in this book, we could revisit the vehicles of Fig. 4.3 or the scientific database about lyrebirds in Sect. 4.2.2.1. I use an African Wildlife Ontology in my textbook, as it is relevant locally and the rest of the

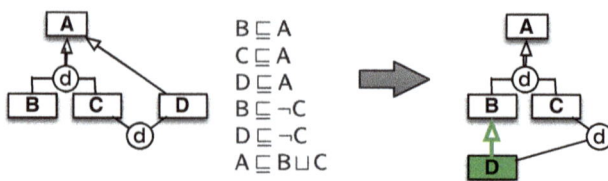

Fig. 5.3 Automated reasoning example before the reasoning, rendered diagrammatically and formalised in a suitable Description Logic (left) and after running the reasoner, with the deductions shown in green (right)

world can daydream about going on a safari. The thing with logic is that we can just as well do without any subject domain and just use A, B, C, ..., since, at the core, it's about the truth value of a statement in relation to the other, regardless of what we think of when we read a particular term. And also for this reasoning task, there are 'negative' examples where things go wrong—like our inconsistent mud house—and 'positive' examples where we obtain desirable deductions—as with Lindiwe's mud house.

A mini-ontology is depicted in Fig. 5.3, with the Description logics axioms side-by-side with abuse of EER notation to visualise them. A could be animal, B carnivore, C herbivore, and D omnivore, say. Whichever example—as long as B and C are subclasses of A and they're plausibly disjoint and making up A, and there's some narrative that there are D's, too, on cursory glance anyway, and that they're definitely not C's.

Subsumption is indicated as per usual with an arrow in a diagram: the rule of inference is that if an object is an instance of the subclass, then it's an instance of the superclass. That is, if $B \sqsubseteq A$ then all the objects that are B's are also A's. Subsumption is transitive, so if there were a subclass $E \sqsubseteq B$, then all instances of E would also be instances of A. In the other direction, downward into the hierarchy, there is something called property inheritance from the parent to the child, which comes implicit with the rule of subsumption: if all B's are A's, then those B's surely have at least the same features as the A's, and additional ones such that they carve out a subset at least in theory.

There's disjointness again in this example, denoted with an encircled "d", as we've seen with the house and mud, the vehicles, and the lyrebird database models. A new feature is the 'complete', indicated with the double shafted arrow, which is formalised as $A \sqsubseteq B \sqcup C$: all A's are either a B or a C or both. Wait, what? Yes, it indeed said "or both": the completeness constraint doesn't care which of them or whether an object is an instance of both. That sounds like it contradicts with the disjointness axiom that no object can be both. It doesn't. The two axioms taken together amount to an exhaustive partitioning of A by B and C, or, stated in a controlled natural language, it can be rendered as "each instance of A must be either a B or a C, but not both". It's like we have seen with conceptual data models in the previous chapter, but then not a 'take my word for it and hope for the best

5.1 What Is an Ontology, the Artefact?

in an implementation'; rather, a 'here's the logic foundation to enforce that in the implementation'.

That's all there is for the declared knowledge. It isn't wrong, but it isn't great either. D appears to be oddly dangling on the side in Fig. 5.3. If all A's must be either a B or a C, then what about D? Can D never have any instances then? Not quite. The algorithms of the automated reasoner gives the benefit of the doubt to what we've declared and it tries to find a way so that D still can have instances and does not contradict the axioms provided as its input—that its instances must be instances of A and they can't be also instances of C. Let's look at the options. D can't be a subclass of C because of the declared disjointness, because if so, it would make D unsatisfiable, which must be avoided. There's nothing that states D can't be a B either. With $D \sqsubseteq B$, it satisfies that it's not a C and thanks to the transitivity of the subsumption, the instances of D are still instances of A, and that for any possible instances we can have with this logical theory. It necessarily must hold, too. We say that $D \sqsubseteq B$ is *derived* from the ones that were already declared.

As a final example, a deduction thanks to properties. We've seen the idea already in the previous chapter on conceptual data modelling: each entity subtype must have at least one addition or stricter feature for it to be a subtype. That holds also here. Figure 5.4 shows a 'sloppy' logical theory on the left-hand side. Again, it's not wrong, but it's not great either. It must be the case in all possible worlds, in all structures of the logical theory, that D's are B's, simply because it has that additional $\exists S.F$ that B doesn't have.

All the deductions we've seen are logically correct, yet maybe you don't want either of them. Maybe D wasn't supposed to be a type of B. To get rid of such an undesirable deduction, the ontology has to be modified somewhere, for which there are several options. For instance, the covering constraint could be removed so that D doesn't have to be a B and so then it remains a sibling class of B. In larger ontologies, further modifications may be feasible so that the deduction won't happen again if you don't want it.

Of course, we can port this feature back into conceptual data modelling, but it's hardly done in industry. Why not is anyone's guess. It allows a modeller to find mistakes early on during development time and therewith produce a better conceptual data model, just like we do during ontology development, and the better model results in a better-quality database.

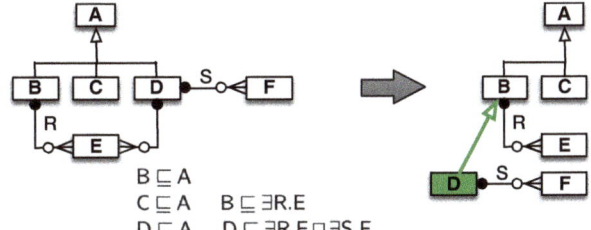

Fig. 5.4 Automated reasoning example showing class subsumption: before (left) and after (right) running the automated reasoner, rendered diagrammatically and formalised in a suitable Description Logic, with the deduction shown in green

$B \sqsubseteq A$
$C \sqsubseteq A$ $B \sqsubseteq \exists R.E$
$D \sqsubseteq A$ $D \sqsubseteq \exists R.E \sqcap \exists S.F$

5.1.3 An Ontology Is More Than Just a Logical Theory

One might be tempted to equate a logical theory with an ontology. They're not the same. Logic is indifferent about your favourite subject domain. A statement may be logically permissible and not contradictory, but be complete nonsense with respect to the subject domain. Like stating that apples are bananas, Apple $\sqsubseteq$ Banana: it's syntactically correct and the reasoner will not flag it as wrong if nothing else in the theory contradicts it. Logic also has no preference whatsoever between an axiom that states that 'for all x's that are sphere-shaped, those x's are oranges' and 'for all x's that are oranges, they have as shape a sphere', as it doesn't between (5.3a) and (5.3b) of the mud houses—but humans do. For some reason, the second option is more adequate than the first and we'd better have that one in the ontology rather than the first one. So, somehow, one ontology may be *better* than another with respect to representing the subject domain accurately. In contrast, logical theories just are.

To be clear, ontologies do take the benefits of logics—precision and automated reasoning. But what they demand as well, are scientific and other theoretical insights that enable one to put forward an argumentation why one way of representing a particular piece of information or knowledge is better than another. And to follow through on that in the ontology. Oranges, the fruits, have a shape that is spherical, not that there are spherical objects with an orange-ness. The shape of an object is a property of that object, not vice versa, and the 'base thingie', also called the bearer, takes precedence over the property that it has. The orange is the bearer that holds the shape, not that the sphere holds the object. There is no 'orange-ness' to a sphere, but there is a 'sphereness' to the fruit. Further, if the orange were to be a bit squeezed into an oval or grown in a cubic-shaped container, it's still an orange and we all would identify it and classify it as one despite the uncommon shape. The spherical shape of the orange is not essential to the orange, whereas the way we identify an orange, is. An analogous argument can be made of the colour of the banana: it's the banana that is the bearer of the green, yellow, and brown, not that those colours are bearing the banana and the one thing to identify an object as a banana. Qualities like shape and colour are disparate kinds of things from physical objects. They are changeable properties of the physical objects, and indistinguishable instances of them can't be used for identity and identifying an object by their very nature of being identical. Two objects, on the other hand, like two bananas, can be identified as bananas, whereas two instances of a yellow don't tell us anything of what the yellow entities are. These examples are informal descriptions of the philosophical arguments underlying them which hint at the explanation why the two options are not the same. To dig into that, we'd need to take a leaf out of the next chapter. Assuming for a moment that those details are correct and since we should want to represent the reality, or our best understanding thereof, in an ontology, we'd better choose the more plausible axiom of the two, which is the one where all oranges have a spherical shape.

There are other instances of modelling patterns where one of them is theoretically better than the alternatives across the subject domain examples. Objects and the roles they play is another popular one. We'd have 'all lawyers are humans', not 'all humans are lawyers'. To justify the preference, the argument of rigidity of properties can be used or lifespan with the temporal aspect of the objects. In layperson terms, we know from evidence that most humans are not born lawyers—no one, in fact, since it requires many years of education and passing the bar exam to be let into the profession. So, the lifespan of a lawyer is shorter than that of a human and not all humans are lawyers; hence, it cannot be the second option. To say that logic doesn't care may be pushing it too far: this can be formalised in a suitable logic. Most logics in use, however, are not expressive enough and so I end up repeating a long story about this each time I teach ontology engineering, as without the expressive logic, it's up to the modeller to not make that mistake. After that, we can push it further with categories of entities. For instance, entities like lawyers, teachers, and football players are *roles* that people as *objects* play. The human as physical object is the bearer of those roles, or the roles inhere in a physical object. Or: we end up with a precedence ordering here as well, and emanating from that, guidance on which way of modelling the knowledge is better ontologically. These sort of insights, be it with roles or shapes or colours, aren't part and parcel of the logic layer itself, but on top of it, as it were. We need the logic to be precise about it, but there's that 'more' to ontologies than solely the relations of truth values *between* one and the other statement: it's also the value *of* the statement, on what it is supposed to be with respect to reality.

5.2 Success Stories of Using Ontologies

Now that we have gone through theoretical aspects of ontologies: what does all that give us? Ontologies have come to the rescue in multiple applications, or measurably improved the results compared to not using an ontology in the tool. It's hard to choose from the myriad of examples. Ontologies directing data integration for digital humanities research, where archaeologists and food historians are trying to find out the food network in the Roman Empire to understand the robustness of trade and food security better. Ontologies for finding six Long Covid clusters thanks to semantic integration of electronic health record from 38 sources using the Human Phenotype Ontology. Or to save millions of euros on expensive geologists trying to find places where there may be oil or gas in the ground still. And on making strides in genomics research and scientific knowledge discovery where the automated reasoner beats the human experts. Or medicine, ecology, agriculture, e-government, or music, to name but a few subject domains.[7]

---

[7] Respectively: the food application (Calvanese et al. 2016), the long covid clusters (Reese et al. 2023), the Statoil use case (Kharlamov et al. 2017). The one in genomics and on outperforming the human experts are described next. For medicine, there are numerous examples as well, with several of them linked to SNOMED CT (SNOMED CT 2023). An early example on the benefit of ontologies in ecology is presented by Madin et al. (2008).

Since there are several ways how an ontology can be useful in software systems, I selected the success stories such that each one demonstrates a different category of ways of using ontologies in information systems. We start with the most widely used application of ontologies, being for data integration and interoperability, and then proceed to benefits of automated reasoning where the deductions surpassed the humans. They're among the expected ways of how to use an ontology. Then there are those who jumped on the ontologies bandwagon and invented various other purposes for them. We'll explore a humble application in e-learning, even though it just as well could have been chatbots or Google's search engine. All examples aren't easily bested by large language models, if ever.

5.2.1 Data Integration With the Gene Ontology

Ontologies were proposed in the mid 1990s to solve the data integration problem, positioning an ontology as a unifying layer on top of those conceptual data models for the separate applications that had to be integrated. The computer scientists were busy trying to figure out how exactly that was supposed to work, both technically with respect to the actual integration and on building and deploying the ontologies. The geneticists and, moreover, the bioinformaticians supporting them, had a serious data integration problem and were scouting for solutions. They stumbled upon this new wonder potion, that ontologies somehow would solve all their data integration problems. How exactly wasn't fully clear to them at the time, but they felt they couldn't wait on the computer scientists for too long.

Scientists and bioinformaticians from the Flybase, a database about fruit flies (*Drosophila melanogaster*), from the Saccharomyces Genome Database, on baker's yeast, and those from the Mouse Genome Database got together in 1997, trying to figure out what that 'ontology' thing was and how it might solve their problems. They knew that parts of the discovered components and mechanisms of the cell and the molecular biology of these model organisms were similar or even the same across species, but everyone was using their own terminology and storing their own data in isolated data silos. It was even a contest of sorts to come up with the funniest names of genes—sonic hedgehog for spike production in fruit flies, the casanova gene in zebrafish that's needed for proper heart formation, the Ken and Barbie genes and Grim and Reaper genes, and many other fanciful names that mix pop culture with whatever it is that the gene product is doing or the process it is interfering with in that species. A good number of them exist in other species as well and those geneticists more often than not would have given another fanciful name to the gene and gene product. But what if you could combine the knowledge or glean hints from one of the other model organisms to speed up your own research for your favourite species? That only would be possible if there were to be a *lingua franca* for the names of the genes, the description of the function of the protein that the gene encoded for, and which processes it was involved in. That shared, structured, vocabulary had to be developed. The ideas were solidified in 1998 with the creation of the Gene Ontology Consortium, work commenced, and the first big coming out

5.2 Success Stories of Using Ontologies   95

of the ideas and results was reported in the prestigious *Nature* journal in 2000.[8] Interestingly, they didn't do it quite in the way that the early ontology papers were aiming for. It worked nonetheless.

The Web Ontology Language, OWL, didn't exist yet—mutually incompatible ontology languages did—and the logicians were only just about making a breakthrough in efficient automated reasoning that had not yet trickled down to the end user stage at that time. The first key aspect was that the bioinformaticians of those model organism databases added their own ontology language, abbreviated with OBO. The three distinguishing characteristics of the language are that it's a directed acyclic graph, uses opaque identifiers, like a GO0002022, with mandatory labels as a means for natural language separation, and has two core relations, being 'is a' and 'part of'. The second aspect was that 'the GO' are actually three structured controlled vocabularies, as they humbly characterised their artefacts: one ontology about molecular functions, one about cellular components, and one about biological processes, and in such a way as to work for any species, regardless of whether they are bacteria, insects, humans and what have you. Third, the approach to data integration differs from the way how the integration was supposed to take place, or, well, unlike what the computer scientists had been investigating.

That integration is at the data layer, rather than the information layer it had been envisioned for. Not just that, they do it in a way that can make a novice ontologist's head hurt. There are things at the type level stored in OBO format in the Gene Ontology, that are then recast as values for database use due to some transformation script, that are eventually shown to the domain experts as if they are entity types. The domain expert is spared these convoluted conversions at the back-end and thinks of it all the same. There were two reasons to convert the OBO into a relational database format: performance and the actual integration use. Instead of linking the ontology to a conceptual data model, they link the terms in the ontology to individual rows in multiple databases. The set-up is illustrated in Fig. 5.5 and elaborated on in the remainder of this paragraph. Take, for instance, the Kyoto Encyclopedia of Genes and Genomes (KEGG) database set up in Kyoto, Japan, since 1995, which stores data to assist with investigating "high-level functions and utilities of the biological system, such as the cell, the organism and the ecosystem, from molecular-level information"; as of September 2022, it has over 42 million genes indexed, 552 metabolic pathways, nearly 19,000 metabolites, and then some for 8383 organisms.[9] The KEGG database has entries in the database identified with meaningless identifiers, like K01834, each with its own name given to it, like gpmA, its own other fields that the developers deemed of interest, and a field in a database table column called "other databases", which contains identifiers to other systems, including the Gene Ontology. For K01834, this includes GO0004619

---

[8] That first article (Gene Ontology Consortium 2000) set in motion a snowball effect for ontologies in related subject domains, a lot of use, and tools to support the ontologies.

[9] The database is accessible at https://www.genome.jp/kegg/ (last accessed on 29-5-2023) and the first key reference paper is (Kanehisa and Goto 2000).

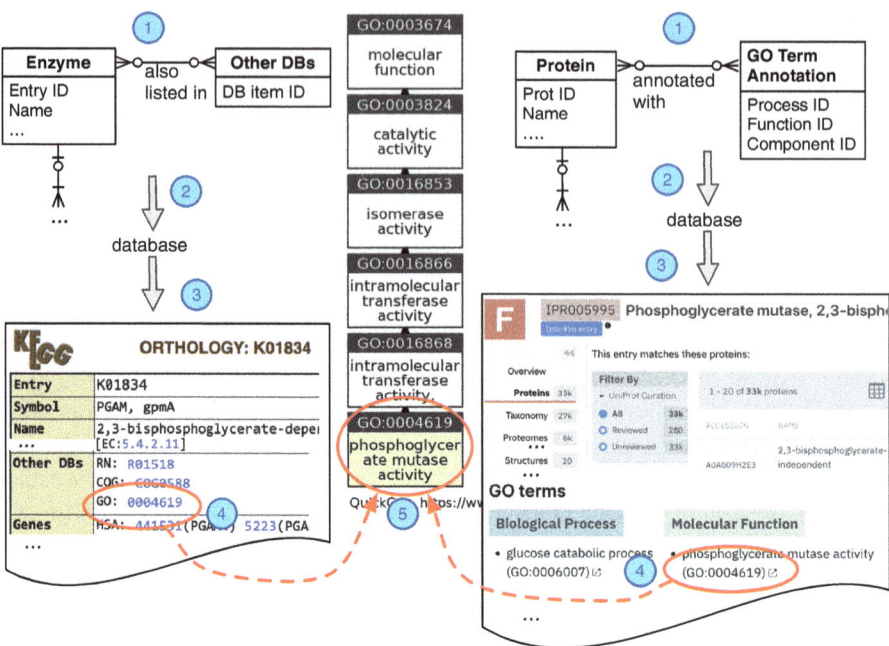

Fig. 5.5 Sketch of an example of data-level data integration with the Gene Ontology: (1) each organisation designs their own conceptual data model, that (2) is converted into a database where data about enzymes, proteins, genes etc. are stored. Each entry, such as KEGG's K01834, is annotated with a Gene Ontology entity (e.g., GO0004619), among other data in the database. All information about each enzyme (resp. protein, gene etc.) is then rendered on a web page (3), where annotation terms are hyperlinked (4) to an entity in the Gene Ontology (5). Conclusion: KEGG's K01834 and InterPro's IPR005995 talk about the same thing, being GO0004619 in the Gene Ontology, which provides the common vocabulary across databases

"phosphyglycerate mutase activity", pointing to an entity in the Gene Ontology. When the data from the database table is rendered on a webpage, a prefix is added to annotations like GO0004619, so that those identifiers are shown has hyperlinks that simplify navigation directly to the linked record in that other database, without even knowing it is really another database that may well be physically stored on a machine standing in an air-conditioned server room at the other end of the world. Consider now the InterPro database's record identified with IPR005995. The developers of InterPro, located in the UK, focus on the functional analysis of proteins rather than metabolic processes.[10] Its IPR005995 entry is annotated with GO0004619 as well. Nonetheless their different emphasis, and thus also the additional data it stores about IPR005995 in the InterPro database compared to

---

[10] The InterPro database integrates 13 protein signature databases with additional annotations and is housed on the infrastructure of the European Bioinformatics Institute in the UK, at https://www.ebi.ac.uk/interpro/ (last accessed on 29-5-2023); see also Blum et al. (2020).

K01834 in KEGG, both deemed it of use to annotate these entries with a term of the Gene Ontology to indicate what it does. Since they link to the same entity on the Gene Ontology, they assert that they store data about the same thing. The database itself stores only the identifiers of the external sources, but when each entity is rendered on their own webpage, the external identifier is prefixed with the respective external database prefix, such as "http://amigo.geneontology.org/amigo/term/", so as to generate a clickable hyperlink pointing to http://amigo.geneontology.org/amigo/term/GO:0004619, being the web page for that Gene Ontology entity. Any database can link to other database entries like that—distinct local identifiers, complementary information—thanks to using the Gene Ontology terms that help with figuring out such agreements across databases. And so on for many, many other databases, linking database record by database record. Currently, there are 33 contributing members in the Gene Ontology Consortium, covering database also about Arabidopsis and amoebae, fish and flatworm, and more for specialised resources to other model organism databases. Our GO0004619 has 324 annotations as of May 2023, which are all listed on that webpage functioning as central pivot. It connects between, among others, UniProt, PomBase, Reactome, MGI and other three-letter-acronyms for species as diverse as human, cow, walnut, zebrafish, and garden lettuce that all have a gene product with the GO0004619 phosphyglycerate mutase activity molecular function.

Nearly twenty-five years since its inception, it's the most comprehensive resource of structured knowledge on gene products, by a large margin. The Gene Ontology has been cited by over 100,000 scientific publications, there are about 8 million annotations with the nearly 44,000 terms in the ontology for 1.5 million annotated gene products of some 5000 species. Collaborations are increasing, which facilitate extensions of coverage and quality, as is uptake of new techniques and tool development and use for new features. The latter now also extends to annotations that link Gene Ontology annotations, using their "causal activity modelling" approach. It's a bit like with the mixed acid fermentation we saw in Chap. 3, but then emphasising the arrows (enzymes that are the gene products) and making a computable version of it. Indeed, such techniques are available. Those diagrams from Chap. 3 can be made fit for the twenty-first century.[11]

As mentioned, the Gene Ontology doesn't use the originally envisioned way of integration of the database schemas or their respective conceptual data models, but its approach of referring to rows in database tables of other databases is highly effective. Its effectiveness made other people note it and made it branch out from data integration to, among others, more efficient literature research and data mining and knowledge discovery.[12] Also noteworthy is that the Gene Ontology

---

[11] Recent developments are described in (Gene Ontology Consortium 2021). The latest release statistics are available at http://geneontology.org/stats.html. Uptake of new techniques and languages includes Shape Expressions (ShEx) (https://shex.io/ (last accessed on 29-5-2023)). Causal activity modelling with GO-CAM was introduced in (Thomas et al. 2019).

[12] Classifying PubMed literature by Gene Ontology terms in their hierarchical position was possible in the first version of GoPubMed in 2005 (Doms and Schroeder 2005). One of the multiple

ontologies are, technically, directed acyclic graphs, not unlike the graphs of the currently popular *knowledge graphs*, but then contributing to open science and being accessible to anyone who'd like to have a look at it and use it.

5.2.2 Outperforming the Scientists and Engineers

It's the stuff of AI movies and hyped-up online media: the machine beats the human in a non-trivial task. Not the variety of tasks of recognising a cat in a picture, but reasoning to a conclusion based on arguments inputted. This can be automated; in fact, it's been done for many years. When I mentioned that feature during a brief radio interview in early 2022 on the ontology of pandemic, the interviewer, John Gericke of SAfm Sunrise, was apprehensive of even the idea to relegate reasoning to the machines. He is not the only one with such a view. Many a novice ontologist or domain expert runs a few test with the automated reasoner to check that it deduces what it should deduce. The standard reasoners for ontologies make the implicitly represented knowledge explicit; there's no extrapolating or approximating or statistical probabilities. We might have to set up a charm offensive for the public to clarify that the automated reasoners for ontologies aren't going to take over the world. They can be used for a specific set of tasks in specialised domains—and can do it well. It's a modest step forward.

It has been shown achievable for automated reasoning over ontologies in the 2000s already, some 15 years ago from the time I'm writing this section. Not with massive amounts of data and electricity-guzzling algorithms running on clusters of big machines as the current machine learning and deep learning approaches require, but with the scientists' knowledge, a language to represent it, and an automated reasoner that can run on a single machine.

It started out as a use case for answering the question: can ontologies represented in the Web Ontology Language OWL together with the automated reasoner derive something *interesting*? Katy Wolstencroft, then a PhD student at the University of Manchester, UK, was tasked with trying to answer it. The use case was in the subject domain of protein phosphatases, which are a certain type of enzymes that remove phosphate from some molecule. That may sound harmless and simple, but there are several ways to do that with varying specificity and in different species. They also have something to do with diabetes and cancer. The phosphatases family of enzymes has sub-families, and within sub-families, there are further variations of the different components the molecules have, how many of them, their functions, and which molecules they interact with. Such details were formally represented in the

---

examples with the Gene Ontology and data mining is described as early as 2002 (Perez-Iratxeta et al. 2002) where GO terms and mining are used to associate diseases ('phenotype') with gene products.

5.2 Success Stories of Using Ontologies

ontology. An example is relegated to the footnote;[13] here, I'll substitute those long and obscure names with recognisable shapes to facilitate reading and understanding. For instance, for the R2A phosphatase subfamily, it should state somewhere that it has: (1) two [squares] and one [rectangle] (both because its parent class has those), (2) four [triangles], and then (3) one of each a [circle], [oval], and [diamond]. What matters is that other types of phosphatases have variations on this theme, like no [oval] but three [circles] rather, or none of that but only one [hexagon]. The human experts represented and classified the proteins manually; the reasoner did so automatically.

The first key outcome of the automated reasoning was that it detected two additional shapes (functional domains) that the human experts had overlooked in their categorisation such that the human categorisation could be refined. Or: it outperformed the humans, largely because at some point it got too complicated and too much for a few humans to memorise it all, whereas the size of the ontology can simply increase. Also, this structured documentation was then used to bootstrap annotations for protein phosphatases of another species, being the fungus *Aspergillus fumigatus* that infects especially immunocompromised individuals, as well as to compare the differences across species. Along the way of the automated classification, it found a phosphatase with a domain architecture that did not fit any known ones, or: it derived something that was even newer to the humans. About the calcineurin protein Afu5g09360, if you must know, which has an additional homeobox domain.[14]

A second use case was also useful for scientists, but then in a wholly incomparable way. Alessandro Mosca and Matteo Palmonari, then with the University of Milano-Bicocca in Italy, had a collaboration with a company that makes tyres. Pirelli, to be precise, who also sponsor sports, including the Inter Milan soccer club and they supply the tyres for Formula 1 racing. As mundane an activity that tyre production may sound like, it hardly is. Not all tyres are alike, and researchers investigate the materials to create tyres of various desirable properties. The difference between summer and winter tyres but the most common ones. The brute force approach of trial and error—try any combination of chemicals and production processes in the lab and see what happens—is far too expensive. Their key question was about finding a way to reduce the theoretical possibilities of hundreds of options to a handful of the most likely combinations of compounds to meet the specified optimisation goal.

---

[13] That is: for the R2A phosphatase subfamily, it should state somewhere that it has: (1) two protein tyrosine phosphatase p-domains and one transmembrane p-domain (both because its parent class has those), (2) four fibronectin p-domains (some part of the molecule), and then (3) one of each a immunoglobulin p-domain, MAM p-domain, and cadherin-like p-domain. What matters is that other types of phosphatases have variations on this theme, like no MAM p-domain but three immunoglobulin p-domains rather, or none of any of all that but only one adhesion recognition site.

[14] Details from a molecular biology perspective are described in (Wolstencroft et al. 2006) and from the ontologies and system development perspective in (Stevens et al. 2007).

The interesting aspect about this second use case is that they turned the automated reasoner upside-down. They were one of the earlier ones to take this comparatively new opportunity of automated reasoning with instance classification and did something with it that was not intended and not foreseen by the logicians—and yet it worked out well. Instead of trying to make sure everything classified without errors as it is supposed to be, they let the reasoner run and the more objects that showed up as inconsistent the merrier. The inconsistent ones for sure did not have to be investigated in the lab, and thus amounted to a bigger reduction of the possibilities, therewith saving the company a lot of money.[15] Thus, also in this example, ontologies assisted human reasoning, when the subject domain had become too large for one or a few humans to oversee and process it all.

5.2.3 Automatic Question Generation and Marking with Ontologies

There are things that are used for a task to solve a problem that was unanticipated during design. That does not only hold for medicinal accidents, like Viagra from angina medication, and for use-function inventions, like opening a beer bottle with a lighter or using scissors for screwdriver. Software can be repurposed as well, such as blogging software ending up being used increasingly as website software instead (Wordpress). Another example is ontologies. Of the many examples, I'll illustrate the unexpected use of ontologies for e-Learning to create intelligent textbooks.

Online versions of textbooks need not be just the pdf of the hardcopy. But what else can you do with a softcopy textbook? While that question was being raised, the pressures on teachers and academics kept, and still keep, on increasing due to more students being enrolled that are needier and due to expanding lists of administrative tasks, without the commensurate increase in staff. This led to explorations into new ways to automate teaching processes. Manually creating homework exercises and tests and marking them is time-consuming because it's a knowledge-intensive task. Ontologies contain knowledge. Instead of only storing knowledge in an ontology, one might as well try to extract some of that out of it. An axiom, say, Lion $\sqsubseteq$ ∃eats.Impala can be verbalised as a statement, 'each lion eats some impala', and as a question, like 'does each lion eat at least one impala?' or 'each lion eats some impala. true or false?' or 'are impalas eaten by lions?' or 'which of the following animals do lions eat? (a) butterfly (b) impala (c) spider'. With that one axiom in the ontology, the answer will be yes, true, yes, and b, respectively, which also comes straight from the ontology. This obviously can be turned up a notch or two for more complex axioms and varied questions and embedded in an e-learning system.

From the engineering viewpoint, you need to know which axiom types go with which question types, how to make the question look grammatically and orthographically correct, and have a suitable question set generator. The first such

---

[15] The details are described by Mosca and Palmonari (2007).

system was developed by Vinay Chaudhri and colleagues from the company SRI, based in California, USA. They developed an ontology in the subject domain of cell biology, which was the topic of the textbook chosen for the use case. They then linked the classes and properties of the ontology to the term usage in the text of the softcopy version of the textbook. A pretty front-end interface has visual cues that a term in the text has a question associated to it. For instance, about the mitochondrion. The ontology contains, for instance, the axiom for 'Each mitochondrion has some ribosome as part and has some matrix as part and has some granule as part'. The system can then generate answerable questions like "Can ribosomes be part of the mitochondrion? Y/N" and "Mitochondria have which of the following things as part? Select all that apply. (a) ribosome, (b) granule, (c) distractor 1, (d) distractor 2", where the distractors are other terms from the ontology's vocabulary that are not asserted to be part of mitochondria. The answers to the questions are computed from the knowledge declared in the ontology. Generating multiple choice questions from ontologies also has received ample attention as well as a larger set of typical educational questions. Moreover, to do so with algorithms that, ideally, can handle any ontology, or at least a greater number than tailor-made systems for a single ontology.[16]

A solution to one problem—data integration—can be, and indeed has been, used to solve another problem.

5.2.4 Ontologies as the Panacea?

To sum up what we've seen so far in the light of solving the limitations of conceptual data models, ontologies can fix those issues from the previous chapter: the lack of precision, lack of detecting certain quality issues of inconsistency, and the data integration challenge. Semantic variation points are stamped out by using a logic. There's a plethora of logics each with their own small, medium, or large set of features. While the 'lightweight' languages such as the OWL 2 profiles have very few features, which can be used as argument that precision is found wanting, that's an incomparable category of imprecision: of the language features permitted in a language, we know exactly what they mean, unlike in UML. The use of a logic, and, more precisely, one that is computationally usable, helps to solve the quality problem: the automated reasoner can find unsatisfiable classes that otherwise may go undetected and find implicit constraints that may or may not be intended but are spotted at design time regardless. There are also assistive methods and tools to get the modelling right so that what is represented and inferred is useful for, among others, science and education. (More about that shortly.)

The main reason for ontologies at their inception was to be instrumental in solving the database and software integration problem. We've seen that with the

---

[16] The system by SRI was developed around 2010-2014 by Chaudhri et al. (2013). See (Kurdi et al. 2020; Raboanary et al. 2021) for the broadening of questions and reusability.

Gene Ontology for data-level integration across records in multiple databases, which has been remarkably successful. The other scenario—the one that the computer scientists had in mind but clearly did not communicate well—is the type-level or schema-level integration. The ontology as sitting on top of a collection of conceptual data models of multiple applications or one large and complex system, and those conceptual model elements all link up to the content of the ontology. Meanwhile, ontologies are used for that as well, but given the amount of uptake of ontologies for other tasks it is, perhaps, the least-often used one.

Ontologies also fix problems we didn't know we had at the time ontologies were the new kid on the block. Yet, the panacea, or silver bullet, to any software issue ontologies are not; there are limitations. But let's first look at how to build one to reap the benefits of ontologies.

5.3 Methodologies for Developing Ontologies

In contrast to the dearth of methods and methodologies for developing mind maps, biological models, and conceptual models, there are tons of them for ontologies. A greater number than you may wish for. Why I don't know for sure. Developing a good conceptual data model is easier, or less hard, than developing a good ontology, largely because there are fewer things to take into consideration. The scope is broader for ontologies, a larger amount of people tend to be involved, and numerous non-ontological resources may have to be consulted. You can do it top-down, bottom-up, middle-out, follow a scenario, use a waterfall sequence, iterate with that waterfall, be agile, dress up test-driven strategies, and more.

Waterfall methodologies are outdated, but they offer the conceptually simplest approach. It's a straw to grasp and hold on to, consisting of a set of basic steps whose guidelines are meant to be broken down into smaller sub-steps and extended. It goes roughly as follows, generalising from several such waterfall-based approaches:

1. Conduct a feasibility study. What are the problems to solve (and assess whether an ontology would really be the solution), opportunities, potential solutions that may exist or to develop with the prospective ontology, and whether developing an ontology is economically feasible.
2. Carry out a domain analysis. Devise motivating scenarios for what you think you want to do if only you had that ontology already. Specify so-called competency questions, which the prospective ontology itself must contain the answers to, and therewith demarcate the scope for the ontology.
3. Plan and sketch a design (also called the 'conceptualisation stage'). This contains one or more of the following activities, among others: look for some of those informal diagrams we have seen in the previous chapters or maybe you have a database from which information may be extracted to serve as draft content for the ontology, search for other ontologies that you could reuse in whole or in part, decide on a structure of the ontology as artefact (e.g., whether it should be one file or consist of modules) and the content's backbone hierarchy.

4. Ontology authoring. That is, 'implement' it by adding the content to the ontology, suitably formalised in a logic of choice, and conduct quality testing along the way to check that you're indeed building what you set out to do. It may be that you get stuck here at some point, and then you'll have to go back to step 2.
5. Deploy! Your ontology unlikely will ever be really finished, but at some point, it will be good enough to release it and do with it what you intended to do with it, be it ontology-based search, integration, negotiation or whatever other task.
6. Maintain it. Not that there will be wear and tear to the ontology, but 'maintenance' in the sense of, among other options: add the extra content that didn't make it into the previous version, adapt the ontology to new requirements, or correct a modelling mistake that may have come to the fore during its use. And create some documentation for posterity; it's easy to forget what you were thinking during development and it will help users outside your developers' group to grasp what's represented in the ontology and why.

This sequence still leaves enough wriggle room for a rhino. For starters, it also allows iteration over whole or part of the sequence and it can be morphed into a lifecycle.[17] Most of the variation comes from the myriad of details, or component-steps, at each step along the way. We're going to colour it in with two approaches, being bottom-up and top-down.

5.3.1 Bottom-Up Approaches to Ontology Development

The 'conceptualisation' phase in step 3 should be appealing by now: to try to squeeze candidate content out of the sort of models we have seen so far, among others. Squeezing semantics out of conceptual models, of thesauri, of biological models, of databases, of spreadsheets, and even out of mind maps and text. There are methods and tools for each of them and they fall under the bottom-up approach to ontology development.[18] The less structured the original non-ontological material is, the harder it is to do and the more manual intervention it requires—under the assumption you'd want to develop a good quality ontology.

Since there are so many biological models already and they look structured, then how can we squeeze candidate content for ontologies out of them? The question has been posed before and I've tried to answer it as well. Your best bet would be on those biological models for which there are tools to draw them. Those tools have a set of icons that are essentially categories or types of things that can be added

---

[17] See Section 5.1 in (Keet 2018) for an overview of ontology development methodologies. An example lifecycle is presented in the 2013 Ontology Summit Communiqué (Neuhaus et al. 2013).

[18] Concrete examples, among many, are: converting a UML class diagram into an OWL file (first reported Gasevic et al. 2004), converting thesauri (Cardillo et al. 2014; Kless et al. 2012; Soergel et al. 2004), databases (Lubyte and Tessaris 2009), spreadsheets (O'Connor et al. 2010), and the brief overview from text, with its sub-tasks, in (Asim et al. 2018).

to the candidate ontology. Once we know what those things are ontologically, we can transfer contents of any such model into an ontology. There are a few extra decisions to make in this bottom-up process, such as which language to use for it, but then it's all set to go. For instance, the PathwayAssist tool has icons for, among others, protein, disease, and cell process, and arrows for, among others, regulation, expression, and molecular transport, which connect the elements. Each such element is then analysed, like that the names ending with '-ases' are enzymes that are proteins performing a function or role, and that proteins are enduring objects (not processes that happen in time). The arrows can be converted into relationships. This analysed base vocabulary of PathwayAssist is then transferred into the ontology and formalised. For instance, each cell process is a process that is located in the cell: **CellProcess** ⊑ **Process** ⊓ ∃located_in.Cell. Given an actual PathwayAssist diagram that has such a yellow rectangular cell process icon, which is labelled with a name, say protein degradation, then that part of the diagram will end up as **ProteinDegradation** ⊑ **CellProcess** in the ontology, that, in turn, deduces that protein degradation is happening within the cell, since its parent does so. After converting many diagrams, it may be crowded among the siblings of all types of cell processes, which will require human intervention to impose any further structure, but at any rate there is that basic axiomatisation. If some other diagram would have put protein degradation as an extra-cellular process or mislabelled a purple rectangle (for disease) with that name, the reasoner will detect the inconsistencies across the diagrams. A human may be able to do so for a few small diagrams, but likely overlook that when it's scaled up to a diagram with hundreds of elements or across tens of diagrams. The ontology-based approach may be your only feasible option.[19]

Also part of the planning and sketching (step 3, still) is that harmless-sounding "maybe there are other ontologies". Maybe you are lucky and there is one that you can use off-the-shelf. There are more subject domains than that there are ontologies, so it would be luck indeed. This raises new questions to answer and methods for how to find the ontology you may reuse in part or in whole. This can be split up into sub-problems to solve. And methods and tools for solving each sub-problem. There's research into how to find a suitable domain ontology and how to find and choose a suitable upper or foundational ontology, and research into how to be able to extract only that part of an existing ontology that's relevant to your domain of interest, for which there are several categories of algorithms and within each category, one algorithm performs better than another in some way.[20] And so on and so forth.

We're not going through all the options here. For each sub-task, eventually, a 'winner' emerges in or for community that may be scientifically the best one, or the one whose proposers had the best connections, or the one with a marketing budget, or the one that was easily understandable to the average user or had the easiest to

---

[19] My answer, the DiDOn method, can be found in (Keet 2012).

[20] See Alharbi et al. (2021) on reuse and Khan and Keet (2015) for a framework on ontology module creation and management.

use tool that was compatible with the most systems or that was open source and for free. Uptake is a different story.

5.3.2 Top-Down Approaches to Ontology Development

A top-down approach can also be taken for ontologies, as we already could for conceptual data models in the previous chapter. Here, however, it needs to reach out to content of the next chapter alike a cherry-picking and mashing it up to something usable by modellers and software. Top-down starts from basic, core, principles of structuring things and to take it from there to descend into the domain ontology minutiae. How does this work? We've already touched upon it: entities have passed the revue that were called objects, processes, functions, or roles. They are general kinds of entities independent of the subject domain. We can put them in an ontology as well! Indeed, that has already happened. These ontologies are called foundational or top-level ontologies, whose developers aimed to make them reusable for domain ontologies in any subject domain. Ontologies that are aligned to the foundational ontology are expected to be compatible with each other. An early well-known foundational ontology is the Descriptive Ontology for Linguistic and Cognitive Engineering, DOLCE (*dolce* means 'sweet' in Italian), that was developed under the leadership of Nicola Guarino at the Laboratory for Applied Ontology in Trento, Italy. It's the same town where the Council of Trent was held in the sixteenth century, as main claim to fame of the small provincial town at the foot of the Alps and worth a short visit. The DOLCE authors put it out in the open solidly on December 31st, 2003. I was to visit the lab in early 2004 for a few months, as an aspiring PhD student, and wanted to learn more about it. It was early days for foundational ontologies and new ones were developed or improved upon over time. Meanwhile, depending on how you count, about 3 to 13 other foundational ontologies have been proposed. While from afar they may look all the same, they do disagree on the fundamentals of what kind of things we have at the top in the hierarchy—each one of them claiming to contain the relevant top-level entities in the right way. The top two layers of the class hierarchy of two of them are depicted in Fig. 5.6, together with a few illustrative examples of the kind of things that the uncommon terms refer to. The one that has the most uptake to date is the logically lightweight top-level ontology Basic Formal Ontology (BFO), which drives a "foundry" of tens of domain ontologies that hang underneath it.[21]

So there comes the question of how to select the best one and then how to align one's domain entities to it. Taking a step back, it requires figuring out what 'best' means, and a step back from that is to figure out what the criteria have to be for determining that and what the values are for each of those foundational ontologies.

---

[21] BFO (Arp et al. 2015), GFO (Herre 2010), DOLCE (Masolo et al. 2003), and others, such as Yamato (Mizoguchi 2010) and UFO (Guizzardi et al. 2022). The 'foundry' with BFO was introduced by Smith et al. (2007).

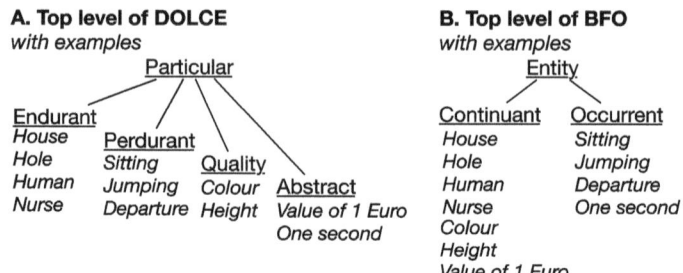

Fig. 5.6 Top-level entities in two of the foundational ontologies that aim to help structuring the domain content for a prospective domain ontology, accompanied by a few examples for each entity

I had proposed an honours (4th-years) project about it in 2011 and Zubeida Khan took it and ran with it, whilst we were at the University of KwaZulu-Natal, in South Africa. She assessed and compared three foundational ontologies on their philosophy and engineering properties and implemented the feature matrix in an ontology selection tool with a set of multiple-choice questions to answer. Given the answers and any optional weighting furnished by the user, it computes the most appropriate foundational ontology for the scenario, which gives a winner or a draw. It then will report which foundational ontology is the most appropriate and why, or that none of them meets all the requirements, and why. The sort of questions have to do with philosophical underpinnings as well as with concrete practicalities. The former is about principles such as whether the ontology should be realist and whether multiple entities can be co-located, like the vase and the amount of clay it's made off. The other group of questions are pedestrian in comparison. Like whether the ontology is represented in the language of preference—if it isn't, it'll cause extra hassle to convert—and down to the sharing licence put onto it.[22] The latter never used to be a problem—anything online was public and free to reuse and modify—but the climate is changing to policing and control and having to use legalese text to state the implied purpose of an ontology: use and reuse.

The foundational ontologies publicly accessible are allowed to be reused. Once one is selected, there's the task of alignment. Is, e.g., our House a so-called Endurant or is it an Occurrent, or something else? There are methods and tools to assist the modeller figure it out, with advances in techniques made as recent as 2022.[23] Be that as it may, just like for mind maps and conceptual data modelling, there are people who make their life's work out of making the mechanisms of development of ontologies better and making ontologies better. It may sound over-

---

[22] The tool is called ONSET, from ONtology Selection and Explanation Tool (Khan and Keet 2012).
[23] (Emeruem et al. 2022).

5.3.3 A Dance Ontology

It's time to revisit the running theme: models about dance. Typically, ontologies are developed in teams over months to years and described in articles taking up at least 10–15 pages to describe the design rationale, the processes followed, the modelling challenges encountered, and solutions created, methods and tools used, some form of evaluation, and how and where it is or can be used. I've done so before as well, albeit largely motivated by the conviction that I do need to get my hands dirty every now and then to experience what it's like and note how annoying the gaps in techniques can be so as to prioritise which one to improve first.[24] I'll present only a small sampling of such a development process in this section.

The first questions to consider are: what do we have on dance ontologies or the like, and what do we need? A quick online search will offer a few ontologies about dance and we have those non-ontological models from the previous chapters that may be amenable to a bottom-up ontology development approach. On closer inspection, they're all about something different: disparate dance topics in the mind map, lyrebird music and dance shows in the biological model, data for the lyrebird shows in the conceptual data model, which salsa and cha-cha moves are taught at which level in the Latin dance ontology, and other movements in the DanceOWL ontology. To bring structure into this hodgepodge, if it's usable at all, we need some methodological approach. Consulting the rudimentary waterfall model from the beginning of this section, it must be admitted that purely dance is hardly economically viable, as it's a challenge with any form of art and even more so with intangible cultural heritage. Let's ignore that for the moment. We then have to decide whether we're going to develop an ontology for the sake of it or for a plausible purpose. Purposes may be annotating dance videos to be able to easily find them or to manage lesson plans at arts colleges, or to combine the two for distance education, or documenting new dance moves. Let's go with the video annotation for online lessons.

As scenario, we take the pupil's need to annotate video fragments and find them. A sample competency question that the ontology will need to have an answer for may be "What are the basic salsa moves for beginners?" and "What is the first step in the Suzy Q move?", and "Do we clap hands in Salsa?". This indicates that the ontology needs to have content about salsa, dance levels, dance moves, and components of dance moves of both the feet and the arms and hands. Looking for ontologies to reuse that may already have at part of that content, there appears to be one that may be reused: Kouthar Dollie, then an honours student at the

---

[24] Some of the ontologies I was involved in include the data mining optimization domain ontology (Keet et al. 2015a) and the African wildlife tutorial ontologies (Keet 2020).

Competency question: "What are the salsa moves for beginners?"

Part of the ontology serialised in OWL functional style syntax

```
...
# Class: :Back_Spot_Turn (:Back_Spot_Turn)
SubClassOf(:Back_Spot_Turn :DanceMove)
SubClassOf(:Back_Spot_Turn
    ObjectSomeValuesFrom(:memberOfDance :Cha-Cha-Cha))
SubClassOf(:Back_Spot_Turn
    ObjectSomeValuesFrom(:memberOfDance :Salsa))
SubClassOf(:Back_Spot_Turn ObjectSomeValuesFrom(:partOf :Level_I))
SubClassOf(:Back_Spot_Turn ObjectSomeValuesFrom(:partOf :Silver))
...
```

DL query over the ontology: partOf some Level_I and memberOfDance some Salsa

Query results: (30 moves and actions, among others:)

```
...
Back_Break
Back_Rock
Back_Spot_Turn
Back_Step
Backward_Slide
Basic_step
Catwalk
...
```

Fig. 5.7 Example of a conceivable sequence steps: from requirements to modelling to represent subject domain content in the draft ontology, to querying and obtaining the results

University of Cape Town, scraped text from dance school websites and used natural language processing to induce a basic vocabulary with candidate classes and object properties (relations). Terms were harmonised, such as merging spelling variants, a few constraints added, and that was mostly it.[25] It can answer the competency question on moves, of which an example is illustrated in Fig. 5.7. It doesn't have all the details of all the moves and one certainly can call into question the naming and use of the two object properties, **partOf** and **memberOfDance**.

Overall, we can reuse this ontology, brush it up, and add content. Additional possibly useable content may be gleaned from DanceOWL. It has content on clapping hands that we may be able to reuse: its **Clap** is an **Action** that is a **Movement** and it is equivalent to **Touch ⊓ ∃isActedBy.Hand ⊓ ∃hasDynamics.StrongAccent**. We may add to the ontology that a hand clap is *part* of a movement and that it's an action that can happen in a move in the Rueda, which is a particular type of salsa dance that is danced in pairs in groups. The Suzy Q move has a right foot cross and a left foot cross, each with a syncopation in the cadence of "1-and-2, 3-and-4". This raises the question of other syncopation patterns and, perhaps, synonyms of syncopation or a description thereof, like 'double-step', that uses the half-beat in-between. If there

---

[25] This is described in more detail in Dollie's honours project report that is available from the project website at https://projects.cs.uct.ac.za/honsproj/cgi-bin/view/2020/dollie_joseph.zip/ (last accessed on 29-5-2023). The OWL file of the Latin Dance ontology is available from http://www.meteck.org/files/ontologies/.

5.3 Methodologies for Developing Ontologies

were no delimiters to the prospective content, you could walk into the alley of beats and beats per minute. And with different dances for distinctive beat patterns, we can drag in the lyrebird shows from Chap. 3. It escalates quickly.

The reason why scope, purpose, and competency questions should be specified for ontologies is to not go off into such tangents. There needs to be a minimum viable product and additional extensions can be added in a next iteration and release. Like we've seen with the sizes and shapes of mind maps, also for ontologies there's no golden number for how many classes, properties, and axioms an ontology should have to say it's large enough for a release. If it can answer the competency questions and it seems to work for the scenario specified, it may be time to move on to the next stage: quality.

Several tools assist with quality control of ontologies, starting with the automated reasoner to check that there are no undesirable deductions or, worse, that the ontology is inconsistent; both need to be fixed first. Other easy tests concern 'niceties', such as consistent naming of the entities, no orphan classes or object properties that are not used after all, and that each element is accompanied by a descriptive annotation. More demanding are checks for disjointness axioms between sibling classes and whether the right part-whole relation has been used. For the salsa dance ontology, it's tempting to add a disjointness axiom between **Move** and **Action**, but an action is a move as well, which thus would render **Action** unsatisfiable, unless the meaning of either is narrowed down to domain-specific terminology and modelled accordingly. For instance, that a **salsa dance move** is a movement that happens in one 8-count and an **salsa action** is a movement that is part of a move occurring during typically 1 or 0.5 beat of an 8-count. Either way, the interaction between the two has to be declared formally.[26]

To increase the ontology's compatibility with other ontologies, an option is to align it to a foundational ontology, which, in doing so, may assist in structuring the content better. There are guidelines for that, too.[27] More generally on the methods and tools for ontology engineering, it's regrettable to admit that it's not clear how many of those checks and methods and tools need to be used for the ontology to be passable for publication. Iterating over a few is certainly a good plan, since they have been shown to result in better quality ontologies. The salsa ontology of Kouthar Dollie is basic, but appeared to be enough to assist annotation and search of videos. For other purposes, such as recording salsa or grasping the details of movements with its actions, a lot of additional content would be needed, in the direction of the DanceOWL but without its data properties.

---

[26] A tool for the 'niceties' is, e.g., the OOPS! tool by Poveda-Villalón et al. (2012). Disjointness declarations can be checked and improved upon with Advocatus diaboli (Ferré and Rudolph 2012) and the part-whole relations with OntoPartS (Keet et al. 2012). There is also a TIPS to prevent common modelling mistakes (Keet et al. 2015b), among many aids.

[27] Notably the DOLCE Decision Diagram D3 (Keet et al. 2013) and the BFO Classifier (Emeruem et al. 2022).

More can be done, as always. This is not a book about just ontologies, however, and we need to move on to their shortcomings and find solutions for them.

5.4 Limitations

It looks easy to develop an ontology on a rainy Sunday afternoon; it's not. This doesn't stop overconfident novices to upload OWL files to a server and make it available online. Not just that, for tasks like document navigation, even an artefact that tries to pass for a scruffy ontology will offer better results than navigation without an ontology. Even a flaky knowledge graph as low-hanging fruit for an ontology does better in Web-based information retrieval than no such structured data. Ask Google; aren't you pleased by their "knowledge panel" that's powered by that sort of structured knowledge and information? But low quality can be an issue in information retrieval, especially when retrieving data. I looked up those knowledge panels for scientists as preparation for a talk on bias in ontologies I presented in early 2022, as I had seen mine changed after I had written two books. Given the context of the talk, I had searched for German computer scientists, of which Google listed only few; other German computer scientists were tagged as being author or researcher or artificial intelligence scientist. Any query answer for "German computer scientists" specifically thus will be incomplete.[28] It's also opaque what drives their classification. Imprecision also can yield too many answers. Conversely, too much precision and analysis paralysis do not move the world forward either. Where the optimal trade-off of rigour, detail, and improvement on the task lies, remains to be seen.

Another recurring complaint is that there are high start-up costs to learn about ontologies and to develop a good quality ontology. On top of that, it's not sexy front-end material to show off, but merely an indispensable back-end artefact to make things work better for the front-end. That's a hard sell. Inherent limitations of ontologies they're not; they are externalities to cause it to likely not ever make it the cool kid on the block according to the masses.

There are inherent limitations to ontologies. This is partly due to the logics and computational complexity, or: the theoretical limitations of what a computer can do. A more powerful machine can reduce the time an automated reasoner takes, but for some logics, the computation may never finish, or it will in theory but the time it will take is beyond anyone's patience. Also for that reason you'll see few, if any, ontologies that take time and temporal features seriously in the logic. The nice benefits that come with automated classification has as the other side of the coin that it limits the language features of the logic, which, in turn, limits what can

---

[28] The talk was entitled "A preliminary assessment of bias in ontologies—their sources and some possible consequences" and held at the Competence Center on Explainability, Fairness and Acceptability of Intelligent Systems (EFA), University of Ulm, Germany, on 23 February 2022.

be represented in the ontology and what can be deduced. Put differently: a certain amount of imprecision is unavoidable. So, one can complain they're (too) imprecise.

Related to implementations is the unfortunate tool rot. It's not an issue inherent in ontologies, but a broader issue in the sciences. Computer science is trying to grow up concerning reproducibility and reuse, like any other discipline, and ontology engineering is not better than the rest. Most proof-of-concept tools for ontology development are brittle, which does not help uptake.

Another factor is resource reality. In the beginnings in the late 1990s and early 2000s when the sky seemed the limit, ontologies were supposed to be developed for the sake of it. Meanwhile, practically, they're intended to serve at least one purpose, and that mostly in IT and computing for some subject domain. This generally means that an ontology eventually has to be usable on the computer that comes with aforementioned limitations. A further downside of purpose is that if the representation of the knowledge is too close to one application purpose, it may not work for another application, and therewith defeating the purpose of ontologies proper.

It's not quite the end of the road yet, even if it may feel like it. There's one last station to try to address at least the modelling limitations of ontologies, one that doesn't have the issues with computation and size the way ontologies do. It blissfully ignores them, in fact. What is to be found there is the topic of the next chapter.

References

Alharbi R, Tamma V, Grasso F (2021) Characterising the gap between theory and practice of ontology reuse. In: Proceedings of the 11th on Knowledge Capture Conference, K-CAP '21. Association for Computing Machinery, New York, pp 217–224

Arp R, Smith B, Spear AD (2015) Building ontologies with basic formal ontology. The MIT Press, Cambridge

Asim MN, Wasim M, Khan MUG, Mahmood W, Abbasi HM (2018) A survey of ontology learning techniques and applications. Database 2018:bay101

Baader F, Calvanese D, McGuinness DL, Nardi D, Patel-Schneider PF (eds) (2008) The Description Logics Handbook—Theory and Applications, 2nd edn. Cambridge University Press, Cambridge

Blum M, Chang H, Chuguransky S, Grego T, Kandasaamy S, Mitchell A, Nuka G, Paysan-Lafosse T, Qureshi M, Raj S, Richardson L, Salazar GA, Williams L, Bork P, Bridge A, Gough J, Haft DH, Letunic I, Marchler-Bauer A, Mi H, Natale DA, Necci M, Orengo CA, Pandurangan AP, Rivoire CJA C Sigrist, Sillitoe I, Thanki N, Thomas P, Tosatto SCE, Wu CH, Bateman A, Finn RD (2020) The InterPro protein families and domains database: 20 years on. Nucleic Acids Res 49(D1):D344–D354

Calvanese D, Liuzzo P, Mosca A, Remesal J, Rezk M, Rull G (2016) Ontology-based data integration in epnet: production and distribution of food during the roman empire. Eng Appl Artif Intell 51:212–229

Cardillo E, Folino A, Trunfio R, Guarasci R (2014) Towards the reuse of standardized thesauri into ontologies. In: Workshop on Ontology and Semantic Web Patterns (WOP'14), CEUR-WS, vol 1302, pp 26–37

Chaudhri V, Cheng B, Overholtzer A, Roschelle J, Spaulding A, Clark P, Greaves M, Gunning D (2013) Inquire biology: a textbook that answers questions. AI Mag 34(3):55–72

Doms A, Schroeder M (2005) Gopubmed: exploring pubmed with the gene ontology. Nucleic Acids Res 33(suppl2):W783–W786

Dooley D, Hsiao W (2019) 3D visualization of application ontology class hierarchies. In: Melone RS, Hinterwaldner I, Borgo S, Kutz O (eds) The Shape of Things (SHAPES 5.0), Proceedings of JOWO 2019, CEUR-WS, vol 2518, pp 1–6

Emeruem C, Keet CM, Khan ZC, Wang S (2022) BFO Classifier: Aligning domain ontologies to BFO. In: Prince Sales T, Hedblom M, Tan H (eds) FOUST-VI: 6th Workshop on Foundational Ontology, part of JOWO'22, CEUR-WS, vol 3249, p 13p, 15–19 August 2022, Jönköping, Sweden

Ferré S, Rudolph S (2012) Advocatus diaboli—exploratory enrichment of ontologies with negative constraints. In: ten Teije A et al (eds) 18th International Conference on Knowledge Engineering and Knowledge Management (EKAW'12). LNAI, vol 7603. Springer, Berlin, pp 42–56

Gasevic D, Djuric D, Devedzic V, Damjanovi V (2004) Converting UML to owl ontologies. In: Proceedings of the 13th International World Wide Web Conference Alternate Track Papers & Posters, Association for Computing Machinery, New York, pp 488–489

Gene Ontology Consortium (2000) Gene Ontology: tool for the unification of biology. Nat Genet 25:25–29

Gene Ontology Consortium (2021) The Gene Ontology resource: enriching a GOld mine. Nucleic Acids Res 49(D1):D325–D334

Guizzardi G, Benevides AB, Fonseca CM, Porello D, Almeida JPA, Sales TP (2022) UFO: unified foundational ontology. Appl Ontol 17(1):167–210

Hedman S (2004) A first course in logic—an introduction to model theory, proof theory, computability, and complexity. Oxford University Press, Oxford

Herre H (2010) General Formal Ontology (GFO): A foundational ontology for conceptual modeling. In: Poli R, Healy M, Kameas A (eds) Theory and Applications of Ontology: Computer Applications. Springer, Heidelberg, chap 14, pp 297–345

Hodges W (2022) Model Theory. In: Zalta EN (ed) The stanford encyclopedia of philosophy, spring 2022 edn. Metaphysics Research Lab, Stanford University

Kanehisa M, Goto S (2000) Kegg: Kyoto encyclopedia of genes and genomes. Nucleic Acids Res 28(D):27–30

Keet CM (2012) Transforming semi-structured life science diagrams into meaningful domain ontologies with DiDOn. J Biomed Inform 45:482–494

Keet CM (2018) An introduction to ontology engineering, Computing, vol 20. College Publications, UK, 334p

Keet CM (2020) The African wildlife ontology tutorial ontologies. J Biomed Semantics 11(4):1–11

Keet CM, Fernández-Reyes FC, Morales-González A (2012) Representing mereotopological relations in OWL ontologies with ONTOPARTS. In: Simperl E et al (eds) Proceedings of the 9th Extended Semantic Web Conference (ESWC'12). LNCS, vol 7295. Springer, Berlin, pp 240–254

Keet CM, Khan MT, Ghidini C (2013) Guided ENtity reuse and class Expression geneRATOR. In: Benjamins R, d'Aquin M, Gordon A, Gómez-Pérez JM (eds) Seventh International Conference on Knowledge Capture (K-CAP'13). ACM, New York, p a26

Keet CM, Lawrynowicz A, d'Amato C, Kalousis A, Nguyen P, Palma R, Stevens R, Hilario M (2015a) The data mining optimization ontology. Web Semantics: Science, Services and Agents on the World Wide Web 32:43–53

Keet CM, Suárez-Figueroa MC, Poveda-Villalón M (2015b) Pitfalls in ontologies and tips to prevent them. In: Fred A, Dietz JLG, Liu K, Filipe J (eds) Knowledge Discovery, Knowledge Engineering and Knowledge Management: IC3K 2013 Selected Papers, CCIS, vol 454, Springer, Berlin, pp 115–131

Khan Z, Keet CM (2012) ONSET: Automated foundational ontology selection and explanation. In: ten Teije A, et al. (eds) 18th International Conference on Knowledge Engineering and Knowledge Management (EKAW'12). LNAI, vol 7603. Springer, Berlin, pp 237–251

Khan ZC, Keet CM (2015) An empirically-based framework for ontology modularization. Appl Ontol 10(3–4):171–195

Kharlamov E, Hovland D, Skaeveland MG, Bilidas D, Jiménez-Ruiz E, Xiao G, Soylu A, Lanti D, Rezk M, Zheleznyakov D, Giese M, Lie H, Ioannidis Y, Kotidis Y, Koubarakis M, Waaler A (2017) Ontology based data access in Statoil. Web Semantics: Science, Services and Agents on the World Wide Web 44:3–36

Kless D, Jansen L, Lindenthal J, Wiebensohn J (2012) A method for re-engineering a thesaurus into an ontology. In: Donnelly M, Guizzardi G (eds) Proceedings of the Seventh International Conference on Formal Ontology in Information Systems. IOS Press, pp 133–146

Kurdi G, Leo J, Parsia B, Sattler U, Al-Emari S (2020) A systematic review of automatic question generation for educational purposes. Int J Artif Intell Educ 30(1):121–204

Lembo D, Santarelli V, Savo DF, De Giacomo G (2022) Graphol: a graphical language for ontology modeling equivalent to owl 2. Fut Internet 14(3):78

Lubyte L, Tessaris S (2009) Automated extraction of ontologies wrapping relational data sources. In: Bhowmick SS, Küng J, Wagner R (eds) Proceedings of International Conference on Database and Expert Systems Applications (DEXA'09). Springer, Berlin, pp 128–142

Madin JS, Bowers S, Schildhauer MP, Jones MB (2008) Advancing ecological research with ontologies. Trends Ecol Evol 23(3):159–168

Masolo C, Borgo S, Gangemi A, Guarino N, Oltramari A (2003) Ontology library. WonderWeb Deliverable D18 (ver. 1.0, 31-12-2003). http://wonderweb.semanticweb.org

Mizoguchi R (2010) YAMATO: yet another more advanced top-level ontology. In: Proceedings of the Sixth Australasian Ontology Workshop, CRPIT, Conferences in Research and Practice in Information. ACS, Sydney, pp 1–16

Mosca A, Palmonari M (2007) Action based abox update: an example from the chemical compound formulation. In: Calvanese D, Franconi E, Haarslev V, Lembo D, Motik B, Turhan A, Tessaris S (eds) Proceedings of the 2007 International Workshop on Description Logics (DL2007), CEUR-WS.org, CEUR Workshop Proceedings, vol 250

Motik B, Patel-Schneider PF, Parsia B (2009) OWL 2 web ontology language structural specification and functional-style syntax. W3c recommendation, W3C. http://www.w3.org/TR/owl2-syntax/

Neuhaus F, Vizedom A, Baclawski K, Bennett M, Dean M, Denny M, Grüninger M, Hashemi A, Longstreth T, Obrst L, Ray S, Sriram R, Schneider T, Vegetti M, West M, Yim P (2013) Towards ontology evaluation across the life cycle. Appl Ontol 8(3):179–194

O'Connor MJ, Halaschek-Wiener C, Musen MA (2010) Mapping master: a flexible approach for mapping spreadsheets to OWL. In: Patel-Schneider PF et al (eds) Proceedings of the International Semantic Web Conference 2010 (ISWC'10). LNCS, vol 6497. Springer, Berlin, pp 194–208

Perez-Iratxeta C, Bork P, Andrade M (2002) Association of genes to genetically inherited diseases using data mining. Nat Genet 31:316–319

Poveda-Villalón M, Suárez-Figueroa MC, Gómez-Pérez A (2012) Validating ontologies with OOPS! In: ten Teije A et al (eds) 18th International Conference on Knowledge Engineering and Knowledge Management (EKAW'12). LNAI, vol 7603. Springer, Berlin, pp 267–281

Raboanary T, Wang S, Keet CM (2021) Generating answerable questions from ontologies for educational exercises. In: Garoufallou E, Ovalle-Perandones MA, Vlachidis A (eds) 15th Metadata and Semantics Research Conference (MTSR'21). CCIS, vol 1537. Springer, Berlin, pp 28–40

Reese JT, Blau H, Casiraghi E, Bergquist T, Loomba JJ, Callahan TJ, Laraway B, Antonescu C, Coleman B, Gargano M, Wilkins KJ, Cappelletti L, Fontana T, Ammar N, Antony B, Murali TM, Caufield JH, Karlebach G, McMurry JA, Williams A, Moffitt R, Banerjee J, Solomonides AE, Davis H, Kostka K, Valentini G, Sahner D, Chute CG, Madlock-Brown C, Haendel MA, Robinson PN (2023) Generalisable long covid subtypes: findings from the nih n3c and recover programmes. eBioMedicine 87(104413)

Smith B, Ashburner M, Rosse C, Bard J, Bug W, Ceusters W, Goldberg L, Eilbeck K, Ireland A, Mungall C, OBI Consortium T, Leontis N, Rocca-Serra A, Ruttenberg A, Sansone SA, Shah M, Whetzel P, Lewis S (2007) The OBO Foundry: coordinated evolution of ontologies to support biomedical data integration. Nat Biotechnol 25(11):1251–1255

SNOMED CT (2023). http://www.ihtsdo.org/snomed-ct/. Accessed 29 May 2023

Soergel D, Lauser B, Liang A, Fisseha F, Keizer J, Katz S (2004) Reengineering thesauri for new applications: the AGROVOC example. J Digit Inform 4(4):1–23

Stevens R, Egaña Aranguren M, Wolstencroft K, Sattler U, Drummond N, Horridge M, Rector A (2007) Using owl to model biological knowledge. International J Hum-Comput Stud 65(7):583–594

Thomas P, Hill D, Mi H, et al. (2019) Gene ontology causal activity modeling (go-cam) moves beyond go annotations to structured descriptions of biological functions and systems. Nat Genet 51:1429–1433

Wolstencroft K, Lord P, Tabernero L, Brass A, Stevens R (2006) Protein classification using ontology classification. Bioinformatics 22(14):e530–e538

Ontology—With a Capital O

6

> *If you get your core ontology wrong, errors, misinterpretations and misunderstandings flow ceaselessly from your false model of reality.*
>
> — *Mike Hockney, in* Causation and the Principle of Sufficient Reason
>
> *Ontology is more like a playground than a science.*
>
> — *Richard Rorty (philosopher)*

Ontology—with a capital O, the one that can't be counted and for which there's no plural. It's about the investigation into, and characterisation of, the nature of things. What is a part-of relation exactly? And what is the nature of a relation, of causality, and of a collective? What are holes and wholes? And, perhaps moreover in the context of this book: how is that "modelling"? What is a model?

Trying to find answers to such questions is a bit too far out even for most ontology developers. To a few, it finally will give answers as to what the things really are, or the best approximation of our understanding thereof. Such knowledge can provide theoretical underpinnings for why to model things in one way and not another in ontologies and conceptual data models alike, even if the practitioner doesn't realise it. In fact, I did sneak in hints about ontology in the previous chapters, such as on the meaning of the branches in the mind maps, that being a banana and having the colour yellow are examples of disparate types of properties, and the notion of transitivity as a principle underlying a parthood relation. Ontology has the answers—to these and other modelling tasks.

Ontology, with an 'O' rather than an 'o', distinguishes itself from ontology and ontologies, the count noun. Ontology isn't within the realm of compromises, unlike most ontologists who develop ontologies who have to make concessions here and there for a range of reasons. The ontologists doing Ontology, which is a specialisation in philosophy called analytic philosophy, prefer not to settle to get things working and used. This doesn't mean that it's occupational therapy for a lot that's unfit for jobs in industry. Even though most of it is computationally unusable, those new insights may help the modellers to make better models nonetheless. An ontology developer may look at their works and morph it into something usable, often watering down the theory as usability concession for computation or scalability. It would be good if it were to work the other way around as well, where the philosophers would descend from their high horse, look at what the modellers are struggling with, and come up with a solution what it should be then. There's maybe a handful of those.

I only once ever met a philosopher who had found out I had used some of his insights who was delighted that it was useful elsewhere, in another scientific discipline. That was Dr. Joop Leo, then with the University of Utrecht, the Netherlands. He visited me in Cape Town to explore collaboration. We didn't manage to agree on the nature of relations, but the lively discussions had their function. Another philosopher I've cited so often it's bordering groupie status hasn't even so much as acknowledged my existence the few times we've been in the same small-ish venues, nor did he cite my work. Ah, well. The one-way direction is one of those things it probably is ever going to be, for those philosophers who did look over the fence, have found a door to exit to the other side. There are numerous philosophers who have found a happy home in ontology engineering. Or, as Barry Smith, a professor in philosophy with the State University of New York, Buffalo, once formulated it in a 2006 presentation entitled "Why I am not a Philosopher": they've left the philosophy mothership, and that the branch of applied ontology is doing so anyway, just like psychology did over a century ago.[1]

What we're going to do in this chapter, is to take a stroll toward that mothership and peek inside. We'll look at some of that Ontology (henceforth lower-cased) mostly from an applied ontology side. Ontology has a long history, which helps set the stage, and then we proceed to two examples that bear relevance to ontologies and conceptual data modelling and to biological models. Not to mind maps; that's long past us in this journey along ways of modelling. The only way that mind maps can be dragged into the picture at this stage of modelling, is to draw one for the contents of this chapter, because its imprecision is utterly unsalvageable by design. Here's mine, in Fig. 6.1.

---

[1] A section of the talk by Barry Smith is available on YouTube at https://www.youtube.com/watch?v=e5zjZnbi-ZA (last accessed on 4-6-2023).

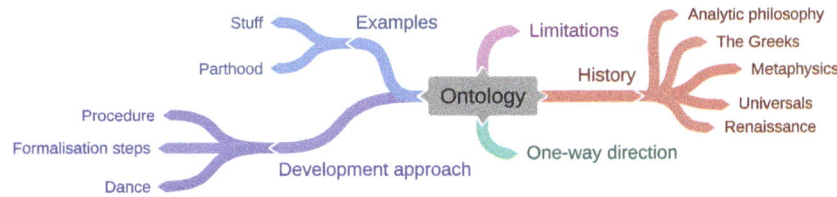

Fig. 6.1 A mind map of this chapter

6.1 The Greeks and Then Some

Where and when is ontology claimed to have originated and evolved to what it is now? These few pages of history on ontology will provide a sketch of it. Illustrious Greeks are credited for starting it all, indeed, but the term 'ontology' was coined only in the seventeenth century in Latin, as *ontologia*. There are several contenders for a claim to fame for coining the term during the time it was still within metaphysics. One is Rudolf Goclenius in 1636, known as Rudolf Göckel before he became a scholar—the Latin renaming a customary practice in those days. Another is Iacobo Lorhardo (AKA Jacob Lorhard) in his Ogdoas Scholastica of 1606 that has *Ontologia* printed on the richly decorated front cover of the schoolbook of the Gymnasium in St. Gallen, Switzerland, where he was the rector. More important for the topic of this book: Lorhardo was influenced by the humanist, logician, and educational reformer Pierre de la Ramée (AKA Petrus Ramus), who is said to have found diagrams indispensable as part of pedagogical tools to structure things to make the core contents of the long and detailed texts more accessible to a wider public.[2]

What they were trying to do in their own way, and revolutionary pedagogy for the times, was to create headings for sections, tables of contents, and figures that look like proto versions of mini-taxonomies and mind maps that fan out from left to right. Structuring material so that it was easier to learn compared to the more time-consuming reading of many books and summarising or memorising those. The material was still based on the classics and still written in Latin, although De la Ramée ventured into writing in the vernacular as well, being French for him. The classics for this topic are chiefly Aristoteles (Aristotle in English) and his teacher

---

[2] The encyclopaedia.com encyclopaedia claims Rudolf Goclenius was first in 1636 (source: https://www.encyclopedia.com/humanities/encyclopedias-almanacs-transcripts-and-maps/ontology-history) but it's more widely credited to Lorhard. An English translation by Sara L Uckelman of the relevant section of his Scholastica (its Chapter 8) is available at https://eprints.illc.uva.nl/id/eprint/668/1/X-2008-04.text.pdf, whereas the whole book has been scanned and is accessible at least via googlebooks and the internet archive https://archive.org/details/bub_gb_rM5gdGMu-rAC/page/n7/mode/2up, if you care to read it in Latin (last accessed on 4-6-2023). Further information about Ramus can be found in the Stanford Encyclopedia of Philosophy entry about him (Sellberg 2020).

Platon (Plato), who, as an aside, was taught by Socrates. Aristotle did not agree with his master Plato, and the differences are not reconcilable. It still can make people debate for hours on end, without them knowing that the debate is about two millennia old. There's Platonism as -ism and one can be a Platonist, one who adheres to Platonism. Plato's main contribution is his 'theory of forms', where the 'forms' are non-physical essences of things (the universals, alike abstract invariants) and where the entities in the physical world are imitations, or instantiations of the ideal form, of those essences (the universals), which counts for all entities. This was too fluffy for Aristotle, as there need not be such higher forms that are separate from space and time. Aristotle was convinced there's the real world with real things. The universal—that what its instances have in common—exist *in* the instances. For instance, the universal **House** that we had in our house ontology example in the previous chapter. Platonism has it that there's some abstract thing outside space and time that characterises it, alike a collection of properties, such as 'has a roof as part', 'has a door', 'made of bricks, wood, concrete, or mud', and so on. Even if we bulldozer all the houses in the world to the ground right now, the universal **House** still exists. An Aristotelian take on it implies that if there's no single instance of the universal, there is no such universal, because the universal can only exist in its instances.[3] Take your pick.

Aristotle went on to analyse a lot and write it up. The main collection of his works, from a philosophical viewpoint, is called Metaphysics, or 'those things that come after the physical world', be this literally, figuratively, or simply on the bookshelf where Aristotle's collected works were put—after the books on physics. There's also the Categories of the Organon, a collection of his writings on logic, which is about the categories of entities there are or that he was convinced of there were at the time. They include categories such as substance (physical particulars/individuals/instances), quantity, time, place, state, and action.[4] And with that, we get in the direction of what is investigated in Ontology and used in those ontologies we've seen in the previous chapter.

Aristotle and Plato were rediscovered in the late Middle Ages and their works contributed to ushering in the renaissance. Recall what was happening with that tree of Porphyry we saw in Chap. 1 in Fig. 1.1, created in the third century AD? The monk was trying to chop it down in the early sixteenth century, in vain. Then there was Ramus structuring things and Lorhard's ontologia. It never went away; on the contrary. From the early 1800s, it starts off in Mitteleuropa with Bernard Bolzano who influenced Franz Brentano and from there, there is an intellectual genealogy to currently living philosophers, as well as from Gottlob Frege from later in the late 1800s to influence the illustrious Vienna Circle and other famous (also by now dead) people, such as Ludwig Wittgenstein and Rudolf Carnap. Each one added their own bits and bytes as extensions, refinements, disagreements, and complete novelties. It is, perhaps, a surprise to know that the ubiquitous parthood relation was not analysed

---

[3] For a long summary, see (Cohen and Reeve 2021).

[4] An introduction to the complete list is freely accessible in (Smith 2020).

6.1 The Greeks and Then Some

and described in detail until Stanisław Leśniewski did it in 1916, in Polish, and so it took another while for the rest of the world to realise it and work on it further during the twentieth century.[5]

As for the obligatory name-dropping for philosophers around the twentieth and twenty-first century, and that also make it into the scientific literature of ontologies, among the first ones to consult would also be Charles S Peirce and Willard Van Orman Quine (mentioned not infrequently in papers), and Mario Bunge (from the Bunge, Wand, Weber, which made it into ontology-driven conceptual modelling). Hilary Putnam's name pops up in a number of places as well, as does Saul Kripke's and several others. Moving on to living philosophers, Achille Varzi may deserve to be added to that list for his contributions on parthood, Peter Gärdenfors for his conceptual spaces, and possibly Kit Fine on relations as well; time will tell whether their works have enough staying power.

That much for the genealogy and academic offspring of philosophers with ideas. A key ingredient of ontology is that it's a user of logic as part of the mechanism of inquiry to try to be precise and infer interesting properties and insights from those premises. The predicate logic and description logics as fragments thereof that we've seen in the previous chapter can be traced back to the Stoics in the third century BCE, also in Greece, which went through the Middle East for improvements, to return to Europe in its medieval era, and renaissance. The budding of modern logic—or: as we know it to be now—was only taken up for development from the nineteenth century CE. For instance, the De Morgan's laws date back to the nineteenth century CE.[6] First order predicate logic is attributed to Gottlob Frege from Germany and Charles Sanders Peirce from the USA, a little over a century ago. Indian logic has a longer history and may be interwoven with ontological inquiry. That it did and does feature not nearly as much as the Greeks deserves further inquiry.

Last, I can't not mention natural language. The relationship status between ontology and natural language would be set to "it's complicated" if they both were entities on social media. And even that status would be an understatement; a bird's eye historical view is hard to achieve. This is not only about how natural language relates to ontology, but also that it may have an ontology of itself, or some ontology is behind it in that language reflects an underlying ontology and thus that the idea of it all exists for a very long time indeed. For instance, there are languages that make a distinction between animate and inanimate objects in the grammar, or consider count nous versus mass nouns: it is tempting to extrapolate such key differences in language to an implicit understanding of the world. Likewise, languages that have roots with variations, such as *-fund-* as root for entities to do with learning and variants that are conceptually close to it, such as *isifundo* 'course', *umfundisi*

---

[5] Leśniewski's original publication: Podstawy ogólnej teorii mnogości I (Foundations of the General Theory of Sets, I). Moscow: Popławski, 1916. (Prace Polskiego Koła Naukowego w Moskwie. Sekcya matematyczno-przyrodnicza, No.2.)

[6] The pair of *not (x or y) = not x and not y* and the *not (x and y) = (not x) or (not y)*.

'teacher', *umfundi* 'student', *-funda* 'learn'. Or the (at most) 23 noun classes in Niger-Congo B languages, where various distinct types of things are classified into different noun classes, such as nouns referring to humans in noun class 1, to long thin objects in noun class 11, and to abstract nouns, such a beauty and humanity, in noun class 14. The people who came up with natural language, might, perhaps, been onto something. Or not. That language interaction with ontology—if at all and if so, however that may be—is a long-standing debate that we won't be able to settle, neither here nor is there consensus about it among academics and practitioners.

What complicates the issues further is whether one assumes there to be a reality to which one has either direct or indirect access to or to be uncertain thereof, and if reality is accepted, whether it exists independently or is constructed through language. Each year when I teach the ontology engineering course, I poll my students about their inclination to one or the other stance. There never has been unanimity for any one of the positions. There has been no agreement on it for millennia, albeit with waves of favouring one stance over the other. We'll let that dimension rest in this chapter.

Overall, one may conclude that it's not all thanks or due to the Greeks. They have laid an investigative foundation, but it really moved on with a sputter in the renaissance and then took off since the Enlightenment period and bloomed of activity in the twentieth and now twenty-first century. Let's have a look at two concrete examples of topics that have received plenty of attention in philosophy as well as elsewhere thanks to their importance: parthood and stuff.

6.2 Examples: Parthood and Stuff

Many examples can be introduced to afford a taste of what philosophers, of the ontologist variety, investigate and how fine-grained the analysis can end up being. I still want it to have some clear relevance to what we've seen in the previous chapters, which reduces the scope considerably. It sets aside popular topics such as causality, events, and space. Two remained eventually: mereology and stuff. The former is directly relevant to what we've come across in passing in the previous two chapters with UML class diagrams and the Gene Ontology: something being a part of something else. The latter cuts across three previous chapters: stuff, or matter, is investigated especially by chemists (e.g., the properties of carbon, of plastics) and by ecologists who try to figure out the flow of matter across the food chain. Conceptual modellers need it for a range of applications in industry anywhere where bulk is processed. And it's relevant for ontologies since it's relevant across multiple subject domains and to guide the conceptual modelling.

6.2.1 Revisiting UML's Aggregation Association and GO's Part-of

In the 'turf wars and truces' section in Chap. 4, the aggregation association of UML class diagrams was introduced. That particular type of association indicates that something is part of something else and implies object behaviour in the software, depending on the constraints specified: that when the whole is destroyed, then so are its parts to be removed from computer memory. The previous chapter's section on the Gene Ontology success story had mentioned in passing that one of the two core relations of the Gene Ontology was part-of. Two distinct modelling settings, same idea, and same level of importance such that it got its own unique element in the modelling language. Something must be going on here.

And indeed it does. The parthood relation is considered the second key relation on par with subsumption. The aficionado might prioritise parthood over subsumption. The first proper theories of parthood weren't introduced until a century ago, however, whereas the classification business goes back at least two millennia. One might argue that parthood is not as easy as classification, but classification wasn't cleared up properly until the late 1980s and to this day many a modeller makes plenty of mistakes. Using parthood correctly in a model is not easy either, but also in this case, the underlying core theory is about as easy.

That underlying theory on parthood is squarely in the court of philosophy, unlike subsumption. Of course, philosophers being philosophers, there are squabbles on a few details, but there seems to be a broad consensus that, ontologically, the finest theory of parthood is the one called General Extensional Mereology. It's built up from the simpler Ground Mereology by adding a few axioms that avoid undesirable consequences.

Ground Mereology takes part-of as a so-called primitive relation: it does not have a definition, but it has three characteristics to describe it: it is a reflexive, antisymmetric, and transitive relation. Reflexive means that an object is part of itself, which is uninteresting here but of use elsewhere later. Antisymmetric means that if x is part of y and y is part of x, then x and y must be the same thing. It's highly suspected that it's this axiom that really causes problems to get Ground Mereology into those ontologies that have to meet the decidability requirement. Transitivity we have seen before in Sect. 5.1: if x is part of y and y is part of z, then x is part of z, which used to be a challenge computationally around the turn of the century, but is doable since. We can stick to natural language descriptions of the theory, or also formalise it, alike we did in the previous chapter. For those who'd like to try it: it's shown in the successive boxed environments. For those who don't: it's very well possible to read the remainder of this section by skipping over the boxes.

---

1. Ground Mereology

(1) $\forall x (p(x, x))$ — parthood is reflexive
(2) $\forall x, y (p(x, y) \land p(y, x) \to x = y)$ — parthood is antisymmetric
(3) $\forall x, y, z (p(x, y) \land p(y, z) \to p(x, z))$ — parthood is transitive

These three axioms may look meagre, but we can play with them already. For instance, now we can define proper parthood in terms of parthood: if x is a proper part of y, then x is a part of y and y is not a part of x, and vice versa.

---

2. Ground Mereology, with Proper Parthood

(4) $\forall x, y (pp(x, y) \leftrightarrow p(x, y) \wedge \neg p(y, x))$ definition of proper parthood

---

With that definition in hand as well, it's possible to *formally prove* properties of proper parthood. Not a proof in the sense of a CSI investigation, a court case, or an experiment in the lab, but using logic, its rules of inference, and one of the proof techniques.[7] The same technique as we already used for reasoning over ontologies in the previous chapter, but then on paper. We can prove that proper parthood is irreflexive, asymmetric, and transitive. Irreflexive means that x is not part of itself, which meets our informal intuition of something being part of something else. Asymmetric means that if x is part of y, then y is not part of x, which also matches a layperson's intuition on parts better. Or: mostly, when you say 'part', like in 'the roof is part of the house', ontologically, the relation you're really referring to is *proper part*, according to the ontologists.

---

3. On Proving Asymmetry of Proper Parthood

The claim is that proper parthood is asymmetric, i.e.,

(5) $\forall x, y (pp(x, y) \rightarrow \neg pp(x, y))$ asymmetric proper parthood

It looks plausible based on the right-hand side of the definition in (4). If so, then it must be the case, in our tableau-based proof, that adding its negation to our theory so far must always result in a contradiction. Negating (5) and rewriting it into negation normal form, it must be that adding

  (i) $\forall x, y (pp(x, y) \wedge pp(y, x))$

to (4) must lead to a contradiction. The next step is to take the bi-implication of (4) as two implications, and start with the first one to see how it fares,

  (ii) $\forall x, y (pp(x, y) \rightarrow p(x, y) \wedge \neg p(y, x))$ first part, from (4)
  (iii) $\forall x, y (\neg pp(x, y) \vee (p(x, y) \wedge \neg p(y, x)))$ rewrite, from (ii)
  (iv) Both $pp(x, y)$ and $pp(y, x)$ must hold   'and' rule, from (i)

(continued)

---

[7] Solow (2005) wrote an accessible broad introduction on how to read and do proofs.

6.2 Examples: Parthood and Stuff

Apply the 'or rule' to create two branches, one for each side of the "∨" in (iii):

(v-a) $\forall x, y \neg pp(x, y)$ first branch splitting (iii)
This contradicts with the $pp(x, y)$ of (iv), hence, a clash.
(v-b) $\forall x, y(p(x, y) \land \neg p(y, x))$ second branch splitting (iii)
(v-b-1) Both $pp(x, y)$ and $\neg pp(y, x)$ must hold   from (v-b)
This $\neg pp(y, x)$ contradicts with the $pp(y, x)$ of (iv), hence, a clash.

Since both branches have a clash and no further rules apply, proceed to the other direction of the bi-implication, which has an analogous proof. QED.

4. On Proving Irreflexivity of Proper Parthood

The claim is that proper parthood is irreflexive, i.e.,

(6) $\forall x(\neg pp(x, x))$ irreflexive proper parthood

The straightforward way to prove it is to simply cite the proof in Halpin's thesis, page A-6 (p203 in the pdf) that proves that *any* asymmetric relation is also irreflexive. Since proper parthood is a relation, it also applies to it, and QED.

For the sake of illustration:

(i) $\forall x, y(\neg pp(x, y) \lor \neg pp(y, x))$ rewrite (5)
(ii) $\neg \forall x(\neg pp(x, x))$ negation of (6)
(iii) $\exists x(pp(x, x))$ rewrite of (ii)
(iv) $pp(a, a)$ existential instantiation of (iii)
(v) $\neg pp(a, a) \lor \neg pp(a, a)$ universal instantiation in (i)
  (vi-a) $\neg pp(a, a)$ apply 'or' rule
    Clashes with (iv), branch ends here.
  (vi-b) $\neg pp(a, a)$ apply 'or' rule
    Clashes with (iv), branch ends here.

Besides proper parthood, other relations can be defined with just the axioms we have seen for parthood and proper parthood. For instance, the notion of 'overlap': if x overlaps with y, then there's some object z and that z is part of both x and y, the part that's not x yet also part of y. Take the crossroads of two roads: there, at the crossing, that part is the z that both roads share. There are more of those relations, like underlap. Not that anyone would want to use that term in a normal conversation, but it codifies a particular sort of relation between items, such as the obvious cases that both your left and your right kidneys are part of your body (in the 'canonical human being') as are the top and the bottom parts of a bikini part of a bikini. We can build up the theory with its formalisation to cover all these cases from just that primitive parthood relation being transitive, reflexive, and antisymmetric. This immediately raises three questions: (1) what other relations can be defined with the theory we have so far? (2) are we pleased with all the deductions that can be made with what we have so far? And (3) should something be added and if so, what will happen if we add another relevant axiom it?

The short answers are that (1) there are other relations but they're not the most interesting right now, (2) you won't be, and we'll get to that next, and (3) lots can be added, and yes you should want to add content. As a first modest step to address the second and third questions, consider for a moment what you think must happen with the 'remainder' of y when x is part of y. There are two options to choose from: either every proper part must be supplemented by another part that's disjoint from the other, which is called 'weak supplementation', or if an object fails to include another among its parts, then there must be a remainder, which is called 'strong supplementation'. Each option can be captured in an axiom and branched off to make additional theories. The former option with weak supplementation amounts to the theory called Minimal Mereology (MM) and the latter is called Extensional Mereology (EM). If you liked weak more than strong, you didn't join the bandwagon.

5. Minimal Mereology

(1)–(3) Our Ground Mereology axioms
(4)–(6) Proper parthood definition (and the properties that entails, including, for completeness: $\forall x, y, z(pp(x, y) \land pp(y, z) \rightarrow pp(x, z))$)
(7) $\forall x, y(o(x, y) \leftrightarrow \exists z(p(z, x) \land p(z, y)))$ overlap
(8) $\forall x, y(pp(x, y) \rightarrow \exists z(p(z, y) \land \neg o(z, x)))$ weak supplementation

That said, EM is not without problems of its own. Objects that have more than one part that are the same proper parts, are identical, which is called the 'extensionality principle'. This means that with the axioms we have in our theory so far, we can't distinguish between, say, a beautiful bouquet of flowers and the same flowers just randomly thrown on a pile, or puzzle pieces in a box and the solved puzzle laying on dinner table. Intuitively, they're not the same time thing even though they have the exact same parts. We pay more for a nice bouquet than for a stack of flowers and see solving the puzzle for an achievement over merely looking at that box of puzzle pieces. So this would need to be fixed for our EM theory so as to approximate better the real world and how we perceive it. We can resolve this shortcoming by adding an axiom to our theory and giving it a new name thanks to that addition. The new theory is called General Extensional Mereology (GEM).[8]

---

6. General Extensional Mereology

(1)–(3) Our Ground Mereology axioms
(4)–(7) Those extras of proper parthood with its properties and overlap
(9) $\forall x, y \neg (p(y,x) \to \exists z(p(z,y) \land \neg o(z,x)))$ strong supplementation
These axioms and definitions make up EM. For GEM, we add:
(10) $\exists x \phi(x) \to \exists z \forall y (o(y,z) \leftrightarrow \exists x(\phi(x) \land o(y,x)))$ unrestricted fusion

---

Don't lose sleep over memorising, or forgetting instantly, all those names of the mereological theories. What is worthwhile to observe and remember, is the game that is being played here: building up ever more comprehensive theories starting from a simple base and exploring its consequences at each step of an extension when one or more axioms are added for a specific reason. With each iteration, the theory is expected to represent reality with increasing precision without also acquiring undesirable properties.

It is tempting to lift the veil of extensions to sort out a smorgasbord of other issues. For instance, on whether parthood can go on only so far to eventually reach a bottom or end to the partonomy where we shall find a smallest indivisible 'atom' or can go on indefinitely, called 'atom-less gunk'. Or how parts behave in time, i.e., to devise a theory for temporal parthood. How do parts and space interact, be it with topology or geometry? Are there types of part-whole relations? Are those theories universal or do they only fit the analytic philosophy that is practised predominantly in Anglo-Saxon countries (i.e., English-speaking countries in The West)?[9]

While the matter stops here for the philosopher, a knowledge engineer or a conceptual data modeller can take these sorts of insights and bring it to ontology development projects and add precision to conceptual data models. Low-hanging fruit to that extent is to finally nail down that "semantic variation point" of UML's aggregation association. For instance, one could give the "shared aggregation" the semantics of part-of and "composite aggregation" the semantics of proper part-of,

---

[8] There are multiple resources on the core mereological theories, their extensions, the multiple possible positions to start from, and debates about each axiom that does, or does not, go in it. A good, and freely accessible resource is the Stanford Encyclopedia of Philosophy entry on mereology by Varzi (2004) and his latest book (2021), simply called "Mereology" and co-authored with A. J. Cotnoir, provides further details.

[9] Atomicity vs gunk is also addressed in (Varzi 2004), types of part-whole relations started with Winston et al. (1987) with an improved taxonomy in (Keet and Artale 2008), and indications of cultural limitations on at least the established types of parthood are presented and discussed by Keet and Khumalo (2018, 2020). There are also many scientific papers on mereotopology, mereogeometry, and temporal parthood, including by myself and co-authors.

or add additional primitives—core elements, like that diamond shape—for other types of part-whole relations or a further refinement for essential and immutable parts, without having to re-invent the wheel to sort it out. It's also possible to slice and dice each mereological theory to make it amenable to computation. If it's not added to the language explicitly, then those insights still can be used for modelling guidance to help explain it to the modeller or to nudge them into the limited set of feasible and ontologically correct options. In short: it can be used to simplify one's modelling life. We tried, even with supporting methods and tools.[10]

6.2.2 What's Lemonade, Really?

Philosophical investigations are aided by examples to illustrate ideas, to extract issues from, and to check a theory against. For this second example about modelling in philosophy, it's lemonade. It was the key running example in a philosophy paper on *portions of stuff*, together with petroleum and cake, that the authors, Maureen Donnelly and Thomas Bittner, both with the State University of New York at Buffalo, were after trying to characterise.[11] Investigating the ontology of lemonade may sound like a true waste of the taxpayers' money. However, it's an example of a kind of entity that is distinct from objects like a laptop, a table, or a pen, and, importantly, its parts are not like the parts of such objects. It is that kind of amorphous entity and, especially, the notion of portions of those kinds of things, that deserve closer scrutiny due to their seemingly different behaviour from countable objects. They are *stuffs* rather than objects. Stuffs are things like lemonade, petroleum, mayonnaise, foam, wood, and water and they are typically named with mass nouns rather than count nouns. Why would one separate them from the other kind of objects? Firstly, stuffs can't be counted: there is no 'one mayonnaise' or 'two mayonnaises'; only amounts of stuff can be counted, like one dollop of mayonnaise and two dollops of mayonnaise or two brands of mayonnaise. If we have an amount of mayonnaise in the bottle and take a portion of it to go with the fries, then that portion is still mayonnaise. The same holds for lemonade. Try to do that with objects referred to with count nouns: if you saw off a part of a tree, you certainly do not obtain a part that is also a tree. Likewise, removing a part from a laptop does not result into two laptops. The disparate behaviours merit consideration whether the two—stuff and object—are categorically different kinds of things.

---

[10] Among the first and persistent efforts of bringing elements of ontology to conceptual modelling are described in the PhD thesis of Giancarlo Guizzardi (Guizzardi 2005), who was with Twente University, the Netherlands, at the time and meanwhile is a professor there. Intricacies with essential parthood were resolved by Artale et al. (2008). Computational trade-offs with M all the way up to the KGEMT mereotopological theory are described in (Keet and Kutz 2017). A method and tool to assist selecting the right part-whole relation for OWL ontologies is described in (Keet et al. 2012).

[11] A longer analysis of portions of lemonade can be found in (Donnelly and Bittner 2009).

What's going on? And, hence: what is lemonade? Let's first clarify that with 'lemonade' I refer to that drink that is made from an often brightly coloured sweet syrupy concentrate to which water is added. That concentrate is dissolved in an amount of water so that the mixture forms a solution. This results in lemonade that is not countable but of which the amount can be measured; it's a stuff. Dividing the amount of lemonade over three glasses, that amount is split into three portions of the same kind of stuff. But this action can't go on forever and therewith raising the notion of 'least portion': that smallest portion of lemonade that still satisfies the properties of being lemonade. Anything smaller than that, it does not have the properties of lemonade anymore, but we get to its constituents: water molecules, (dissolved) sucrose, and molecules that do the colouring and give the flavour, and maybe also molecules of another type that perform the role of preservative of the syrup. If you so insist, we can consider to be in the constituent mix also a selection of drinking water contaminants, such as pesticides, metals, bacteria, and microplastics; it doesn't make a difference to the argument. The key observation here is that the constituents are other things than the whole stuff is. In the case of lemonade, because there's more than one stuff, lemonade is a mixture, as compared to a pure stuff. Pure stuffs have only one type of stuff they're composed of. This sort of idealisation is convenient for analysis and Donnelly, Bitter and other philosophers looking into stuff and portions thereof assume pure stuff exists all over the place. Chemists know better. Even a diamond has trace elements, that is, other stuff that's not carbon, as does 'high purity gold' have other elements.[12] But even for pure stuffs, we end up at a least portion.

The idealisation can be useful to tease out characteristics of what 'stuff' really is and, from there, how portions of stuff relate to each other. A least portion of a pure stuff is assumed to be much smaller than that of a mixed stuff. A pure stuff is necessarily homogeneous, whereas for mixed stuffs that may or may not be the case; lemonade is homogeneous, wood is not. This affects the portion-making task whilst trying to meet that criterion that a portion is the same type of stuff as the whole it was taken from. And, related to that too, the stability of the mixture. A pure stuff may be comparatively stable, like a slab of gold, but an amount of fresh mud in a bucket will naturally separate into sand and water. Other mixtures need tools to separate the components, such as blood in a reagent tube in a centrifuge where the heavier red blood cells will end up at the bottom of the tube.

How to model all this knowledge? As with mereology in the previous section, a few primitive elements are required here as well, and we'll add more as we go along. Let's first sketch the content of the previous paragraphs. For purposes of structuring things and communicating it, we could draw a sketch of the various components and interactions, in one of those conceptual modelling languages notations we're

---

[12] See the Dictionary of Gems and Gemology (Manutchehr-Danai 2009). High purity gold—the most purified one can get—consists of more than 99.999% gold, but where specialist equipment still can find trace amounts of other stuffs, including silver (Ag), lead (Pb), cadmium (Cd), and zinc (Zn).

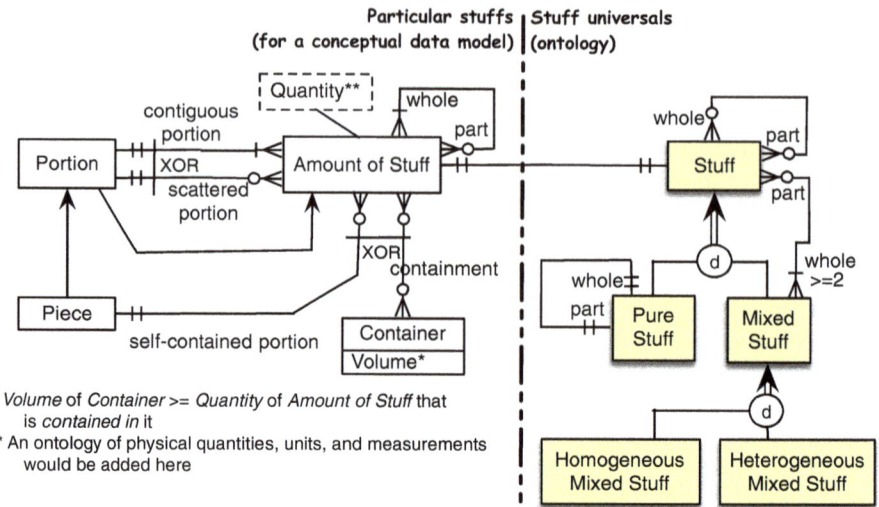

Fig. 6.2 An example of (ab)using EER notation to visualise key aspects of 'stuff': main types and attendant things with their relations, and connecting type-level aspects to classes for instances as individual quantities of the stuff. (Springer copyright image adapted from Keet 2016, with permission)

already familiar with now already—if ontologies can abuse the notation, then so can Ontology. The EER-like diagram is shown in Fig. 6.2: on the right-hand-side we have the types of stuff and how they relate to one another and on the left-hand-side the instance-level notions like the quantities and portions. First, there's Stuff, which is distinguishable from Object. Like in the previous chapter, we might formalise it in Description Logics, as Stuff $\sqsubseteq$ ¬Object, or, as a philosopher will prefer, in first order predicate logic, as $\forall x\,(\text{Stuff}(x) \rightarrow \neg\text{Object}(x))$. Second, we need two core relations: one for talking about the portions of the same type of stuff and one to state that stuff is 'made of' other stuffs, like the sugar in the lemonade. For the latter, we might as well reuse the parthood relation from the previous section. The only question is which of the mereological theories. Since stuff doesn't seem to come close to the identity issue of the bunch and bouquet of flowers, let's be conservative and import only Ground Mereology onto our stuff theory. For the former, a portion-of or has-portion relation, the question is whether we could reuse parthood for that and make it a special variant of it, or whether we have to create a brand-new primitive or defined relation. Attempting to reuse parthood would be akin to trying to get away with formalising that 'portion-of is a kind of part-of where the participating entities are stuffs'

$$\forall x,y\,(\text{portionOf}(x,y) \rightarrow \text{partOf}(x,y) \wedge \text{Stuff}(x) \wedge \text{Stuff}(y)) \qquad (6.1)$$

This option entails that portion-of is also transitive, reflexive, and antisymmetric, since part-of is. Transitive seems plausible: if the sip of lemonade you took was a portion of the lemonade from your glass of lemonade that was poured from the carafe of lemonade, then that sip of lemonade was a portion of what used to be in the carafe. A problem is the "was" in that sentence: this sort of portion, just like a slice of the cake and the dollop of mayonnaise from the bottle, is sensitive to time. The top-half of the lemonade in the carafe as portion of the whole amount of lemonade in the carafe is not. If not time, one still can argue in space: the former scenario is one of separate portions and the latter is a contiguous portion. Choices, choices.

The more we scratch the surface, the more features are laid bare, of stuff in general, of how they relate, and, gradually, also of different kinds of stuff. In this case, it's not only about adding content, but we'll also need a more expressive logic than we have seen so far. Not for the temporal aspects, but for the "Stuff(x)" and the "Stuff(y)" in equation (6.1): currently, $x = y$ is even possible due to the reflexivity, which is not something that should be admissible in our model. The formalisation must state somehow that they are distinct entities yet be of the same kind of stuff, like both being amounts of lemonade. It is possible to formalise this, but not in plain first order predicate logic. We need to resort to either second order logic to quantify over classes and make assertions about them or a many-sorted logic to divide the domain of entities to quantify over. And we don't need that only for defining portions. Lemonade is a mixture. What is a mixture? It's a stuff that is composed of two different types of stuff, so also to definition of mixture needs statements about the classes, not just the instances. And so the quest for ever better axioms continues; the interested reader is referred to the citation in the footnote for the formalisation.[13]

There's plenty to investigate about stuffs still, even after formalising the knowledge discussed in the preceding paragraphs. Unlike what we have seen with Ground Mereology and the more expressive mereological theories built from it, there are no settled 'core' theories of stuff yet. Someday, there will be, provided enough philosophers get to keep their job and be allowed time to do research as stipulated in the job contracts of academics. But is it only an academic exercise, to keep philosophers off the street, alike occupational therapy for the gifted? I would contend not. Also these insights can seep into ontologies and ontology-driven conceptual data modelling where it can be used to build better quality ontologies and models in all subject- and application domains where stuffs are used. A concrete example is traceability of ingredients for food safety and preservation. The industry needs to be able to trace the ingredients from your meal back to the batch it was a portion of, to the bulk amount that the batch was a portion from, to which raw materials were mixed to make up that bulk amount, and so on. The better the traceability management, the faster the source can be identified to prevent further food poisoning or food infections. Two infamous examples of the last decade are the *E. coli* in the bean sprouts in Europe in 2011 and the world's worst listeriosis outbreak, caused by the *Listeria monocytogenes* in a type of fermented sausage

---

[13] The Stuff Ontology is introduced in (Keet 2014) and was extended with a formalisation of these relations in (Keet 2016), which make concessions in the implementation.

called polony, which happened in South Africa in 2017–2018 that caused over a 1000 recorded infections and 216 deaths. A similar scenario plays out in the traceability of ingredients of drugs and vaccines, but then with a focus on catching counterfeit medicines.[14]

6.3 How to Do an Ontological Investigation

Every discipline has its own mores for how to conduct an investigation and for the grander narrative in the field on how to look at things. So, does ontology and how the information is presented. For a particular topic, on paper in the broader setting, there is a, what seems to me fairly typical, cadence in those philosophical investigations. One philosopher first writes an article tentatively stating 'there is something new here, I think'. Replies follow, which is a mix of 'oh yes, and there's more to it' and 'no, really, you are so wrong'. Dust settles. Details are worked out with a suitable logic. They are refined and revisited either *ad infinitum* or until there is consensus. For instance, for stuff, the early works can be traced to Dean Zimmerman and David Barnett from the analytic philosophy side, in articles with wonderful titles like "Some Stuffs Are Not Sums of Stuff", and also touching the intertwined topics of mereology and identity. A separate strand of analysis exists by philosophers in the area of philosophy of chemistry and of philosophy of science, who don't mix much with the analytic philosophers.[15] They hashed out enough details for the ontologists and logicians in Artificial Intelligence to start making those philosophical papers more precise.[16] That combination, in turn, was then adapted and tweaked to get it to work with ontologies, culminating in a Stuff Ontology.[17] There doesn't seem to be a route back from the applied insights and the gaps they observed in the philosophical theories back to inform philosophy, as mentioned before.

At the level of a particular investigation, it looks murky to at least some of my students, but the basics look the same as in other disciplines: you arrive at a question, a hypothesis, or a problem that no one has solved before, do your thing and obtain results, discuss them, and finally conclude. For ontology, what hopefully rolls out of such an investigation is what the nature of the entity under investigation is, and with it, what it's not. For instance, what dispositions are, a new insight into the transitivity of parthood, the nature of the relation between portions of stuff, or what the characteristics of a particular domain entity, like money, peace, or a pandemic,

---

[14] Further information on the motivational scenario about food can be found in (Donnelly 2010; Keet 2014, 2016; Solanki and Brewster 2016; Thakur and Donnelly 2010). Numbers of the South African listeriosis outbreak are from National Listeria Incident Management Team (2018). On compound traceability of medicines, consult, e.g., Bhattacharjee et al. (2015).

[15] See (Zimmerman 1997), (Barnett 2004) and (Donnelly and Bittner 2009). From the chemistry perspective, see (Brakel 1986) and (Needham 2011).

[16] For instance, (Davis 2010; Bittner and Donnelly 2007; Donnelly and Bittner 2009).

[17] (Keet 2014, 2016).

are. The challenges are how to arrive at the question and what actually is supposed to be happening at the 'do your thing' stage to obtain results.

6.3.1 A Tentative Procedure

I haven't seen written cookbook instructions that describe how to carry out the steps in detail for ontology. It is tacit knowledge infused in the way a discipline is taught, just like I knew roughly what to do when I did my thesis project in the area of applied philosophy at Wageningen University. But that was not analytic philosophy. Having read many ontology articles and conducting a few such investigations myself, I think it's feasible to distil a procedure at least for applied ontology. With X to be figured out, which could be anything—a 3-dimensional entity, the nature of a relation, a feature such as the colour of objects, the roles people fulfil, causality, secrets, collectives, you name it—carrying out the following steps will get you closer to an answer as to what the entity is:

1. (optional) Consult dictionaries and the like for what they say about X;
2. Do a scientific literature review on X and, if needed when there's little on X, also look up attendant topics for candidate ideas;
3. Criticise the related work for where they fall short, and how, and narrow down the problem/question regarding X;
4. Put forth your view on the matter, by building up the argument step by step, e.g.:
   (a) From informal explanation to a plausible intermediate stage with sketching a solution (be it in an ad hoc notation for illustration or by abusing, say, UML class diagram notation) to a formal characterisation of X, or the aspect of X if the scope was narrowed down.
   (b) From each piece of informal explanation, create the theory one axiom or definition at a time.
   Either of them may involve proofs for logical consequences and they are subject to iterations of looking up additional scientific literature to be able to finalise an axiom or definition.
5. (optional) Evaluate and implement the theory.
6. Discuss where it gave new insight, note any shortcomings, and mention new questions it may generate or problem it doesn't solve yet, and conclude.

As compared to scientific literature I've read in other disciplines, the ontologists are a rather blunt critical lot when they criticise (step 3). It's worse in person at a seminar or workshop, despite knowing it's partially orchestrated with a designated discussant who prepares a critique beforehand. In twenty-first century parlance: bring popcorn.

The formalisation stage in step 4 has room to manoeuvre. Two visualisations of the sub-steps are shown in Figs. 6.3 and 6.4. Among others, you can choose your logic or make one up on the spot. Logic may look daunting for what it is already, and for many it is deemed part of the great beyond. The basics can take

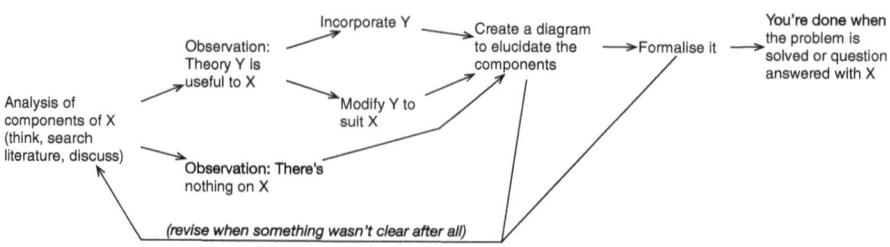

Fig. 6.3 A visualization of the 'iterating over a diagram first' option (4a)

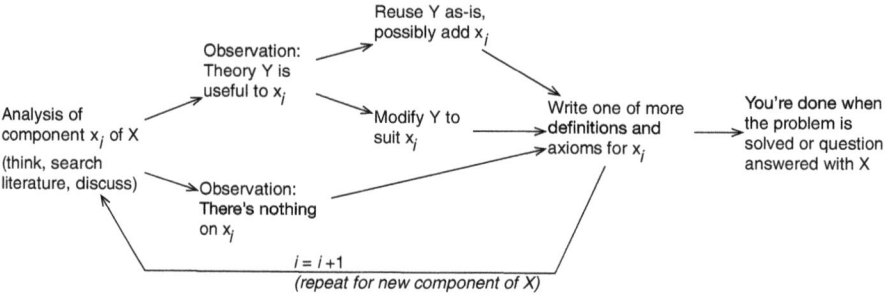

Fig. 6.4 A visualization of the 'one definition or axiom at a time' option (4b)

up a whole textbook as well.[18] Those books present it as a rigid, fixed canon to swallow. However, once those basics are known, it's playtime. You may devise a specialty logic just so or aspire for it to be reused, be it a many-sorted logic defined on the fly, or a second-order logic if the need arises—whatever is needed to capture the semantics unambiguously. Philosophers are mostly users of logic, rather than logicians, and most of them don't bother to encode the ontology into a format that the computer can process to feed it to an automated reasoner. We're to go along with the axioms on paper and hope for the best. I would recommend using it nonetheless, as the tools do exist, such as Isabelle, Prover9, Mace, and HeTS, and it adds credibility to the formalisation.

The other comment on step 4, is that most philosophers are not aware of EER, UML or ORM and use complete freedom in drawing, if there is a drawing at all. Maybe they could be interested, if only the type of model was not called conceptual model, due to the 'concept' in the name, since concepts are mind-dependent entities but, as analytic philosopher, one would want to study the nature of the entity in reality. In a survey among philosophers a few years ago, the overwhelming majority of respondents were realist.[19] Computer scientist and software engineers with their

---

[18] There are thorough introductions to logic that are paper-based (Hedman 2004) or also with software (Barwise and Etchemendy 1993).

[19] The survey aggregates can be found at https://philpapers.org/surveys/results.pl (last accessed on 4-6-2023).

conceptual data models are not all anti-realist, however, but, rather, sloppy in their terminology and lump together class, entity type, object type, or concept, without much, if any, regard of the definitions that philosophers give them. It irked that philosopher about to be proclaiming to leave the mothership, Barry Smith, with IFOMIS at Saarland University at the early years of ontologies, when it caused heated debates. At a bio-ontologies workshop I attended in Rome in 2005, he wanted to train us so badly that he repeatedly proposed to get a jar and order us to put a Euro in it as penalty for each mention of 'concept'. He would have earned himself a fine Italian meal with a good glass of wine or two if we had done so. There are pockets where it may have made a difference, but mostly it reverted to its former sloppy state of term use. Entrenched terminology is hard to change. What it did clear up for the bio-ontologies, is that what goes into an ontology is evidence-based and is indeed grounded on reality, not based on ideology, politics, or profit motives. As for conceptual data modelling, it's within reach to charm philosophers into systematic drawing with UML, EER, ORM, as neither has 'concept' in the language's name nor as term for any of the elements, and we could label it information modelling instead.

Regardless of the 4a and 4b options, what is common for both, is that the philosophical investigations are lonesome endeavours resulting in disproportionally more single-author articles and books than in computer science, biology, or physics. It's also in stark contrast with ontologies: many domain ontologies are developed in teams or even in large consortia, ranging from a few modellers to tens of contributors. They discuss options to get find the underlying cause of the modelling issue, take care of the ontology module on the subject they're expert in, and a team of curators may look at quality across the whole ontology. Whether there's something more to this difference, I don't know.

Is that all there is to it? Roughly, yes. There may be variation in emphasis on one or other particular component of the procedure during an investigation and in the write-up. Different publication venues have different scopes, even if they use the same terminology in their respective call for papers or journal's scope, and that's reflected in a scientific article describing the outcome.

6.3.2 The Ontology of Dance

At last in the series of models about dance—from brainstorming about it with mind maps to dance as done by lyrebirds, to a database about dance, and a simple ontology with terms for salsa video annotation—we get to the question of what dance really is. The familiar may not be as mundane as it initially appears. To illustrate the procedure, let's consider a dictionary and an encyclopaedia. The Merriam-Webster dictionary lists 'dance' as intransitive verb with 2 senses, as transitive verb with 3 senses, and as noun with 5 senses, with the one we've been dealing with before defined as "to move one's body rhythmically usually to music : to engage in or perform a dance".[20] The Encyclopedia Britannica's article on dance has as opening line that dance is "the movement of the body in a rhythmic way, usually to music and within a given space, for the purpose of expressing an idea or emotion, releasing energy, or simply taking delight in the movement itself".[21] They largely agree on the core idea, with the encyclopaedia having a go at the purpose of it, which are essentially additional constraints for something to count as dance. Interestingly, our lyrebird researchers from Chap. 3 had put the bar higher as well: to them, it only counted for an animal to really dance if there are distinguishable dance moves for distinct rhythms, like humans move rhythmically differently on waltz music like the *An der schönen blauen Donau* by Strauss than on a cha-cha-cha on a song like *Oye como va* by Santana. A key question seems to be emerging already: what properties are the necessary and sufficient conditions for 'rhythmic movement of the body' to be deserving to be called dance?

This question gives us a head-start in searching the scientific literature. A search does return hits—people have pondered about the ontology of dance. Not much, but enough to get us a step further again. Two are of particular interest. First, Noël Carroll, a philosophy professor with the City University of New York known for his philosophy on art and aesthetics, took, in his words, a few stabs at the ontology of dance. It charts the landscape of key issues to address, from the obvious as to the nature of dance (i.e., "What is dance?"), to when dance is an art-work, and the always thorny question of identity, *in casu*, when something is a dance as art-work and how to distinguish any two instances from each other. He goes on to state that "Something is a dance only if its choreography contributes indispensably to its constitutive purpose.", which is followed, quite correctly, by his deadpan comment "Needless to say, this formula requires some unpacking." The 'choreography' component of the definition may be the choreography of the ballet of the Swan Lake, the choreographed dances in Beyoncé's music videos, or of the yearly Ladies Ginga Flashmob contest: instructions for how to rhythmically move on the music in a formation alone or with multiple people. A 'constitutive purpose' is what informs the design of the choreography. The rhythmic movements will be different if the purpose is to challenge an opponent for battle versus to seduce onlookers, or when it aims to foster social cohesion.[22]

These purposes Carroll mentions seem to imply premeditated choreography and a clear leaning toward the dance-as-art. It doesn't easily fit with the so-called 'vernacular' dancing, like we do at a wedding or a ballroom or salsa-bachata-kizomba dance social. The lead of the couple has to come up with the choreography on the spot, one move at a time, influenced by the space on the dance floor, theirs and their follow's level of competence, the song, and so on. And it does not exclude marching bands, so it can't be a complete definition. And it doesn't mention there has to be music and, with that omission, that the choreography has to match the

---

[20] https://www.merriam-webster.com/dictionary/dance (last accessed on 4-6-2023).

[21] (Mackrell 2023).

[22] The ontology of dance topics are discussed further by Carroll (2019).

Fig. 6.5 Preliminary diagram on the ontology of dance, first version. The relations that lack the cardinality and mandatory/optional constraints or names, or both, were unclear to me during charting the landscape as part of step 4a of the procedure

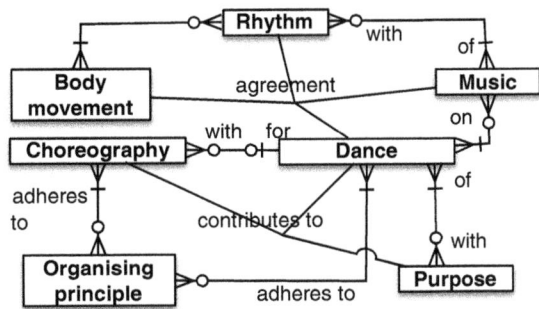

music. Carroll's purposes overlap with those of aforementioned Britannica's article on dance, but it misses especially basic dancer-centred purposes, like the ones on releasing energy and taking delight in the movement. With this paragraph, we've moved from step 2 of the procedure of the ontological investigation to step 3: criticise.

The first complaint, on missing the 'vernacular dance' category of dance, does have a tentative answer in Christian Kronsted's PhD thesis, which he obtained at University of Memphis where he's currently a postdoctoral researcher. His whole PhD thesis is about that category of dance. Ontology was not his aim, but he needed to get the definitional issue out of the way, which he did by offering a list of eight features to describe the "prototypical exemplar of a vernacular dance as meaning-bearing cultural practice at its core". They are: a culturally developed dance system with culturally determined functions and a rich history, adhering to implicit or explicit organizing principles, expressing meaning (virtually or through self-expression) and being more expressive than what is necessary for daily instrumental movements, being performed with music, and containing rhythmic human movement.[23] This rich list of properties has, arguably, one property shining it its absence: purpose. And would a rich history be necessary? That excludes all fad dances and new dance styles, and thus may be too restrictive.

We can continue this game of finding literature and criticising it. Since this is a book about modelling and not dance, we're going to pretend these two papers are enough to move on to step 4 in the procedure: the modelling. Both scientific sources offered neither a diagrammatic representation nor a formalisation, so we'll have to model from scratch. We now can choose option 4a or 4b. Since 4a is more in line with what we've covered in the previous chapters, we'll take that route. A preliminary sketch is shown Fig. 6.5 in ER notation.

Let's commence with the easy aspects. 'Dance' could refer to a (say, salsa) dance at a social whist adhering to its organisational principles, a (salsa) dance performance with a choreography, types of (salsa) dance (e.g., on1, on2, rueda), and that dance (salsa) as a genre. Dance must have organisational principles, but

---

[23] (Kronsted 2021).

need not have a predefined choreography, since social dance with its improvisation is dance as well. The other easy part is that there has to be music, body movement, and rhythm, but how they relate is less evident; perhaps as a quaternary relation. Whether dance has a function or a purpose is harder to answer, for both function and purpose are nontrivial concepts in philosophy. Function concerns the what of the action that the thing is made for, and purpose the why. Do we need a what or a why for dance? I lean toward the why for its increased explanatory power. Which purposes that may be is a topic left for future research. History can be added, but it is even more optional than a choreography. Honestly, I didn't know what to do with the 'culturally developed dance system' and the 'express meaning', because culture and meaning are laden and difficult to define terms. In academese, this is left as an exercise to the reader, or a non-trivial topic for future work.

The next step within the 4a process is to refine this diagram and formalise it, justifying one axiom at a time. We did that partially already, as that's how we got to the ER diagram and so the first move can be to formalise the diagram. That is, all the fully specified relationships in the diagram can be fed into a usual procedure to formalise it, such as the mandatory participation that dance must have a purpose (no matter how frivolous), is done on music, and adheres to at least one organising principle:

$$\forall x\, (Dance(x) \rightarrow \exists y\, (dancedWith(x,y) \wedge Purpose(y))) \tag{6.2}$$

$$\forall x\, (Dance(x) \rightarrow \exists y\, (dancedOn(x,y) \wedge Music(y))) \tag{6.3}$$

$$\forall x\, (Dance(x) \rightarrow \exists y\, (adheresTo(x,y) \wedge OrganisingPrinciple(y))) \tag{6.4}$$

The thorny issues are the two underspecified n-aries: **Choreography** for **Dance** that has a **Purpose** does relate, but it's not clear how, and likewise that there has to be agreement between the **Rhythm** of the **Song** that is danced on during a **Dance** and that the **Body movement** is suited for the rhythm for that song and dance. A bare minimum of formalisation of that can be to merely type the relationship, which means stating there is one and what participates in it, just not how; e.g.,

$$\forall w,x,y,z\, (Agreement(w,x,y,z) \rightarrow Rhythm(w) \wedge Dance(x) \wedge Song(y) \wedge$$
$$BodyMovement(z)) \tag{6.5}$$

At least now it's evident what isn't clear.

Assuming a full formalisation, we then need to evaluate and discuss it. A good starting point is to recall the earlier complaints and to assess whether at least one of them is resolved by the new model. Can, or does, it exclude marching bands? Not in a way I would be able to defend with this model. It fares better for Carroll's choreography: now it's clear that it's not required to have one before dancing the dance, since it must have the more generic organising principles that both a predefined and, if it counts, an on-the-fly choreography must adhere to. The stricter constraint by the lyrebird researchers is suggested as well, with the quaternary agreement relationship that indicates there will be multiple instantiations. Adding an 'at least 2' constraint is an easy addition.

Of course, there are a few loose ends and new questions, as any investigation will uncover. Function and purpose are not the same ontologically, but they were treated as similar enough; perhaps there is a difference when it comes to dance. My ontology of dance requires music, which thus distinguishes it from body movement without it, but not everyone may agree. The 'why' of it wasn't justified convincingly. And, taking that a step further, nothing has yet been asserted about whether the music has to be external to the dances or can be made while dancing, be it like tap dancing or like the lyrebird. Nevertheless, we have made a few steps forward with the ontology of dance.

6.4 Limitations

Yes, also ontology has its limitation for modelling. Mostly, it's too impractical for anyone who wants to put it to use, especially computational use. General Extensional Mereology as a theory may be beautiful and capture the nature of parthood best, but where can that be used? Computationally, hardly. Of course, the theoretician's response would be that there does not need to be any use for it, that just the insight is worthy, for grasping better the nature of things and writing down elegant (logical) theories (informally: models) about them. It is, but a lot of the advances made in ontology are gathering dust. There's a difference between the rest of science and society being late to catch on with some scientific advance, and not bothering to even hand-waive to a potential relevance. Currently, it may well be that at least some of the problems that modellers are struggling with are already solved by philosophers and that word doesn't get out, or not widely enough.

Except, perhaps, for the Stanford Encyclopedia of Philosophy, which is accessible both in terms of readability and free access, and it provides useful comprehensive overviews that may function as a starting place to dig into the ontology literature. That's still only a one-way lane out. There are a few tentative bridges as well, notably with the Applied Ontology journal and the Formal Ontology and Information Systems (FOIS) conference series. I have attended a few FOIS conferences, published in them, and even was the local host of one, published in the Applied Ontology journal and I'm in the editorial board. They aim to bring together philosophers, computer scientists, cognitive scientists, and there may be a linguist or two as well, to foster cross-fertilisation. Yet, I have not seen a flow back into ontology. For instance, they didn't solve the long-standing issue of essential and immutable parts solved; computing came up with a solution, one that is just as impractical to implement as any other model in ontology. Disentangling part-whole relations and sorting out the transitivity issues was neither initiated nor resolved by philosophers; it started in cognitive science and it was enthusiastically taken up by the conceptual data modelling community who was facing concrete modelling issues in need of resolution such that it would not break the software.[24] They are missed opportunities. That said, it happens across other disciplinary boundaries as well, including, admittedly, a myriad of computing solutions to problems that domain experts for bio-ontologies aren't aware of, and there's no easy solution.

This is a book about modelling, not moaning, but it had to be said that indeed also this last stage, in all those steps we took all the way from mind maps to philosophy, there are still limitations. In this case, computation or implementation limitations are insurmountable; the others are of attitude, focus, and disciplinary incentives and customs, and can be overcome in theory if not also in praxis. From here, we can't go any further to fix the limitations to modelling in this way, however. The next chapter will delve into how to deal with it. As spoiler alert: an impasse it is not.

References

Artale A, Guarino N, Keet CM (2008) Formalising temporal constraints on part-whole relations. In: Brewka G, Lang J (eds) 11th International Conference on Principles of Knowledge Representation and Reasoning (KR'08). AAAI Press, pp 673–683

Barnett D (2004) Some stuffs are not sums of stuff. Philos Rev 113(1):89–100

Barwise J, Etchemendy J (1993) The language of first-order logic, 3rd edn. Stanford, USA: CSLI Lecture Notes

Bhattacharjee PS, Solanki M, Bhattacharyya R, Ehrenberg I, Sarma S (2015) VacSeen: a linked data-based information architecture to track vaccines using barcode scan authentication. In: Malone J, Stevens R, Forsberg K, Splendiani A (eds) Proceedings of the 8th International Conference on Semantic Web Applications and Tools for Life Sciences (SWAT4LS'15), CEUR-WS, vol 1546, Cambridge, UK

Bittner T, Donnelly M (2007) A temporal mereology for distinguishing between integral objects and portions of stuff. In: Proceedings of AAAI'07, pp 287–292

Brakel Jv (1986) The chemistry of substances and the philosophy of mass terms. Synthese 69:291–324

Carroll N (2019) Some stabs at the ontology of dance. Midwest Stud Philos 44:70–80

Cohen SM, Reeve CDC (2021) Aristotle's metaphysics. In: Zalta EN (ed) The Stanford Encyclopedia of Philosophy, Winter 2021 edn, Metaphysics Research Lab, Stanford University

Davis E (2010) Ontologies and representations of matter. In: Fox M, Poole D (eds) Proceedings of the Twenty-Fourth AAAI Conference on Artificial Intelligence (AAAI'10). AAAI Press

Donnelly KAM (2010) A short communication - meta data and semantics the industry interface: what does the food industry think are necessary elements for exchange? In: Sánchez-Alonso S, Athanasiadis IN (eds) Metadata and Semantic Research: 4th International Conference, MTSR 2010, pp 131–136

Donnelly M, Bittner T (2009) Summation relations and portions of stuff. Philos Stud 143:167–185

Guizzardi G (2005) Ontological foundations for structural conceptual models. PhD thesis, University of Twente, The Netherlands. Telematica Instituut Fundamental Research Series No. 15

---

[24] The essential and immutable parts can be recast in the temporal dimension (Artale et al. 2008). The part-whole relations in conceptual modelling commenced with Winston et al. (1987) and culminated in (Keet and Artale 2008), where it also becomes evident from the formalisation why some of the 'part of' relations are actually not and not transitive either. Cotnoir and Varzi's recent book on mereology does mention the latter, named mereological pluralism.

Hedman S (2004) A first course in logic—an introduction to model theory, proof theory, computability, and complexity. Oxford University Press, Oxford

Keet CM (2014) A core ontology of macroscopic stuff. In: Janowicz K, Schlobach S (eds) 19th International Conference on Knowledge Engineering and Knowledge Management (EKAW'14). LNAI, vol 8876. Springer, Berlin, pp 209–224

Keet CM (2016) Relating some stuff to other stuff. In: Blomqvist E, Ciancarini P, Poggi F, Vitali F (eds) Proceedings of the 20th International Conference on Knowledge Engineering and Knowledge Management (EKAW'16). LNAI, vol 10024. Springer, Berlin, pp 368–383

Keet CM, Artale A (2008) Representing and reasoning over a taxonomy of part-whole relations. Appl Ontol 3(1–2):91–110

Keet CM, Khumalo L (2018) On the ontology of part-whole relations in Zulu language and culture. In: Borgo S, Hitzler P (eds) 10th International Conference on Formal Ontology in Information Systems 2018 (FOIS'18), vol 306. IOS Press, FAIA, pp 225–238

Keet CM, Khumalo L (2020) Parthood and part—whole relations in Zulu language and culture. Appl Ontol 15(3):361–384

Keet CM, Kutz O (2017) Orchestrating a network of mereo(topo)logical theories. In: Proceedings of the Knowledge Capture Conference (K-CAP'17). ACM, New York, pp 11:1–11:8

Keet CM, Fernández-Reyes FC, Morales-González A (2012) Representing mereotopological relations in OWL ontologies with ONTOPARTS. In: Simperl E et al (eds) Proceedings of the 9th Extended Semantic Web Conference (ESWC'12). LNCS, vol 7295. Springer, Berlin, pp 240–254

Kronsted CSM (2021) An enactivist model of improvisational dance. PhD thesis, Department of Philosophy

Mackrell JR (2023) Dance. In: Encyclopedia Britannica. https://www.britannica.com/art/dance

Manutchehr-Danai M (ed) (2009) Trace elements in diamond. Springer, Berlin, pp 866–866

National Listeria Incident Management Team (2018) Situation report—listeriosis. https://www.nicd.ac.za/wp-content/uploads/2018/07/Listeriosis-outbreak-situation-report-_26July2018_fordistribution.pdf. Accessed 29 May 2023

Needham P (2011) Compounds and mixtures. In: Hendry RF, Needham P, Woody AJ (eds) Handbook of the Philosophy of Science, vol. 6: Philosophy of Chemistry. Elsevier, Amsterdam, pp 271–290

Sellberg E (2020) Petrus Ramus. In: Zalta EN (ed) The Stanford Encyclopedia of Philosophy, Winter 2020 edn, Metaphysics Research Lab, Stanford University

Smith R (2020) Aristotle's Logic. In: Zalta EN (ed) The Stanford Encyclopedia of Philosophy, Fall 2020 edn, Metaphysics Research Lab, Stanford University

Solanki M, Brewster C (2016) OntoPedigree: modelling pedigrees for traceability in supply chains. Semantic Web J 7(5):483–491

Solow D (2005) How to read and do proofs: an introduction to mathematical thought processes, 4th edn. Wiley, Hoboken

Thakur M, Donnelly KAM (2010) Modeling traceability information in soybean value chains. J Food Eng 99:98–105

Varzi AC (2004) Mereology. In: Zalta EN (ed) Stanford Encyclopedia of Philosophy, fall 2004 edn, Stanford. http://plato.stanford.edu/archives/fall2004/entries/mereology/

Winston M, Chaffin R, Herrmann D (1987) A taxonomy of partwhole relations. Cogn Sci 11(4):417–444

Zimmerman DW (1997) Coincident objects: could a 'stuff ontology' help? Analysis 57(1):19–27

Fit For Purpose   7

> *Mirror, mirror on the wall, who's the fairest of them all?*
> — *The Evil Queen (in Snow White)*
>
> *Rational discussion is useful only when there is a significant base of shared assumptions.*
> — *Noam Chomsky*

We've taken a walk along five distinct types of 'models' that fall roughly within the same category: representing information or knowledge that is declarative and a-temporal. Each chapter had a "limitations" section: it's always possible to complain about that way of modelling. Problems, be they real or only when one's goalposts have been moved, are solved elsewhere. But because there are always limitations, does that mean none of them is any good? Of course not. Each one was good in solving at least some, if not all, of the limitations of the other types of models, so they each must be good for something.

In this chapter, we're going to compare them and tease out those details as a way of synthesising the material that was explicitly or implicitly described in the previous chapters. We'll get to an overview of which one to use when, or, put differently: where and how each one is fit for a purpose. Selecting the right type of model will make things run more smoothly.

A second component of reflection is that models can have power, or perhaps more precisely, the people who develop and deploy them, do. There are very many articles on ethics in AI and problems with machine learning and deep learning in applications. Does that apply to these types of models as well? There is remarkably little research on whether it does. The handful of researchers who tried, showed that we cannot rule it out and therefore the second main section of this chapter touches upon it. Finally, taking all this into consideration, you're definitely ready to start exploring designing your own modelling language, which is the topic of the last section of this chapter.

7.1 A Beauty Contest

Just like there are turf wars within a group of models, as we've seen for conceptual data modelling languages, it's conceivable to do so across the various types of models as well. Such across-type blaming, shaming, and praising, however, is typically based on misunderstandings of the other type of models. For instance, complaining about OWL not being good for datatypes, yet not grasping that there shouldn't be any attributes (OWL data properties) with datatypes in an ontology anyway. Or claiming that Ontology is inherently better than conceptual data modelling—think of a 'holier than thou' air of stating something along that line—yet missing the point that those models described in Ontology articles can't be implemented fully anyway. And if the client of the software application really wants to record data violating a theory in philosophy, they can have it nonetheless. It may even be a satisfactory model for their purpose in their information system.

That said, beauty does *not* lie in the eye of the beholder. There are features we can compare each type of model against so as to make an informed decision. Those features have passed the revue in the previous chapters, from their background and aims to their strengths and weaknesses, which are concisely presented in Table 7.1. We shall discuss the feature-based comparison in the next subsection and afterwards turn to two example-based comparisons. The two subsections can be read in either order.

7.1.1 A Feature-Based Comparison of the Types of Models

Let's start with comparing the model types' aims, purposes or function, given that it may need different features to realise them. And, possibly, another type of model can meet an aim or function from another type better. They are summarised in Table 7.1 in the second column for each of the five types of models we've seen. The first observation is that they all do have different purposes.

7.1.1.1 The Type of Model's Aims and Purposes

A particular type of model should satisfy its own aims and declared function at the very least. If it meets any of the others, then that is a coincidental bonus. How well it meets its own aims and how many bonuses a modelling approach and language has, is a non-trivial matter to figure out. How could we know it meets its aim(s) well versus not-so-well versus not really versus not at all? It's a question that already popped up in Chap. 2 on the effects of mind mapping. There are very many variables to test that are not easy to control, ranging from the demographics of the participants, to the methods, tools, and methodologies to develop the model, and the affordances

7.1 A Beauty Contest

Table 7.1 Comparison of types of models along a set of properties

Model type	Feature: Main aim or function	Where used (mainly)	Development methodologies	Software assistance	Language freedom	Precision
Mind Maps	Basic structuring of a topic	Education, business	A little	Yes, many drawing tools	Limited	Low
Biology models	Visualise biology knowledge (structures and processes)	Biology research, textbooks	No	Drawing tools, some runtime usage (simulations)	Ranges from self-imposed to complete freedom	Low/ medium
Conceptual data models	Capture characteristics of data to be stored and processed in an program	Analysis and design stage of database and program development	Yes	Drawing tools, limited runtime usage	Ranges from standardised languages to partial freedom to design one	Medium
Ontologies	Represent knowledge of a subject domain precisely and in a computer processable way	Computing and IT (Data integration, Enterprise systems, Web search, etc.)	Yes, many	Editors (diagram, text), runtime usage	Ranges from standardised languages to partial freedom to design one	High (but medium/ low is possible)
Ontology	Characterise one small aspect of interest precisely and in much detail	Research	No	No	Yes, can define as one goes along	High/ Very high

of the modelling language.[1] As we have seen, ontology development has the most methods, tools, and methodologies, by a very large margin. Then come the aids for conceptual data model development, if it's a numbers game. But do modelling aids and human evaluations thereof that show it's better with them than without imply that the ontology languages aren't good at serving their purpose, since so many aids are needed? Or do those aids merely indicate it's a challenging task to accomplish, noting that difficulty is independent of meeting an aim? Or could we say that based on aids with positive evaluations and the fact that there are many ontologies that are used in a wide range of information systems well beyond their original aim of data integration, they do not only meet their purpose but surpassed it? We don't know. Other measures could be uptake or the percentage of bad models. However, conceptual data models, ontologies, and Ontology were never intended for the masses, nor have they been marketed as such, unlike mind maps and models in biology, so uptake would not be a fair measure for meeting aims. Bad models, like those botched cladograms we have seen in Chap. 3 or a conceptual model in OWL rather than it being an ontology, may be attributed to sub-standard education or newbie enthusiasm rather than a type of model failing to meet its aim. For all these reasons of uncertainty and difficulty of measuring it sufficiently, the 'how well do they meet the stated aims?' is not included in the comparison.

Looking into one model type meeting another model type's aims is easier when considering use cases where cross-usage works out well. Be that 'abusing' or conveniently borrowing one type of model or the modelling language for meeting the aims of another. Yes, you may use mind mapping to start creating an outline for an ontology or draw a conceptual model to sketch the Ontology of dance or whatever topic under investigation, and, yes, you may use an ontology language to develop a conceptual data model if you so insist. These across-type examples do not imply they meet the other type's aim, however. Mind maps fall short on, among others, a precise semantics of relations, i.e., on representing the domain precisely. OWL falls short in language features for conceptual data modelling: it is impossible to represent n-aries fully, and there are no qualified associations nor role join constraints, nor is there a standardised graphical notation for OWL, but if you don't need the missing language features and are not interested in diagrammatic

---

[1] They can be put to the test in experiments. A typical strategy is as follows. First, split a group of participants into two, A and B, be it through random assignment or another way to homogenise the participants. Students tend to be taken as test subjects in academia, and one could order the class according to their rank based on a previous course and then assign the even numbered ranked to set A and odd numbered ones to set B (or vice versa) as a proxy that the groups indeed will be of equal competence. Second, specify one or more tasks to carry out and make group A use the intervention and B not. This may be a selected modelling language for a task, or any of the proposed procedures to design them. Third, compare the models created in a meaningful way, ideally measurable on a scale or against a gold standard, and/or the time it took to develop them. Faster development in the intervention group A and just as good coverage as the baseline group B is a good outcome, as is the same amount of time with a better model; faster and better is even better. The results are often accompanied by a note of caution, in that there may be interfering factors, or the group is too small for drawing firm conclusions. And replication studies are not popular.

renderings of your conceptual data model, then you can safely reuse that ontology language to develop your conceptual data model. Conversely, a graphical conceptual data modelling language also can meet only a fragment of the purposes of Ontology and ontologies, but if the use is within that fragment, it will work. A conceptual data modelling language with a logic-based reconstruction is as precise as the logical theory a model visualises, but your theory should have no need for features beyond what the logic and the icons of the diagram language offer. Conceptual models, by design, have neither a notation for defined classes nor for defined relationships, and EER and UML don't have a notation for any relational property, such as transitivity and irreflexivity, among other missing features. If your ontology doesn't have those either, it's perfectly fine to reuse the notation of EER or UML for your lightweight ontology.

7.1.1.2 Language Freedom and Precision

Let's move on to the next two comparison features in Table 7.1, which are practically inter-related: language freedom and precision. The former concerns whether you can go design your own modelling language to come up with a notation as you go along and whether there is a 'grammar police' for the model like there is for written text. The table's column's values range from 'limited' for mind mapping to 'total freedom' for ontology. The latter, precision, is about the number of features in the language, since the more features available, the more precise one can be in representing the topic. For instance, recall that ER doesn't allow cardinalities greater than one, whereas EER does. So, we can only state, say, 'each car has at least one wheel' in ER, even though we know that cars have 3 or 4 wheels. A database based on the ER model would permit 5 wheels to be declared as parts of a manufactured car, i.e., permit wrong data. The database based on an EER model that has 'each car has between 3 and 4 wheels' is more precise and avoids such dirty data.

Both features tie in with *expressiveness* of the representation language. Expressiveness refers both to the kind things one can state in that language from the viewpoint of modelling and language features and to computational complexity. To complicate matters, they mutually affect each other in different ways. For computation, it's not about the number of features in the language as if it were a bean counting exercise, but which ones. Antisymmetry is computationally costly, so is rigidity on relationships; either one alone in an otherwise simple language may get you into undecidability, which is undesirable for computation because then the reasoner may never terminate. In the context of this book, we use mostly the former usage of the term. A choice for a modelling language with few modelling features means that one not only can, but is forced to, be imprecise. Like selecting a low precision language were disjointness cannot be asserted and so a faulty multiple inheritance of a subclass goes undetected. Lower precision is an enabler of obfuscating disagreement and mistakes and of preventing them from surfacing.

For the models we've covered, their precision runs from 'low' for mind mapping to 'very high' for ontology. Lamentations about the low precision filled the mind mapping limitations in Sect. 2.3: they're just lines and the lines can mean anything—subsumption, parthood, instantiation, examples or exemplars, or attributes, to name common hidden meanings. There isn't a lot one can play with when drawing mind maps, but differences there are. Starting with thick branches radiating from the central concept, to thinner branches, to thinnest twigs versus all lines with the same thickness. A rainbow palette of colours for the branches versus black branches, and capitalised text versus lower-cased text. They are all *presentation decisions* of *the same* elements. Presentation matters—ask any human-computer interaction expert—but now think beyond that boxed-in environment: what if you could add elements? Not just additional lines and extra images, but other kinds of things. For instance, to visually represent a *not* doing something with a red cross at the end at that keyword or overlaid on the branch leading to the keyword. Or allow branches to merge or leaves to differentiate to distinguish processes from objects. The downside of making it less imprecise is that it will take more time to master such a mind mapping language, and that trade-off is hard to avoid for any declarative modelling of information or knowledge. Additional language features for greater precision come at a cost of a steeper learning curve. We've seen that already with the models in biology: there's greater differentiation but if you're not the least bit trained, it's easy to overlook essential subtleties.

Conceptual data modelling languages may be modified to some extent, and they differ on the precision scale. The staple of what we teach to undergraduate students is fixed regardless. They have to learn one specific way of drawing UML class diagrams and EER diagrams and, at times, ORM; a violation will cost them marks in the tests, assignments, and exams. Thou shalt know the rules before breaking them. There is freedom to devise new notations and several notational and expressiveness variants are available for EER and ORM, as well as various extensions, as we have seen in Chap. 4.[2] A controlled reduction is also not off the table. My collaborator Pablo Fillottrani, with the Universidad Nacional Del Sur, Argentina, and I did so for the evidence-based conceptual data modelling languages: include only those features that have been shown to be used most often by modellers.[3]

Language freedom and precision for ontologies is tricky practically. In theory, you do have the freedom; it's just that no-one wants to play with you in your sandbox if you devise a logic that is not compatible with the current ones that have ample software infrastructure. That means mostly OWL, some OBO, and a little Common Logic. OBO has a comparatively low precision and only its designers had the freedom to decide. OWL is a mixed bag: there are actually eight 'species' of OWL, some of which have low expressiveness and others comparatively high and therewith a relatively high precision can be obtained. The OWL Full and OWL 2 Full species allow you to modify the language, but hardly anyone tries intentionally.

---

[2] Such as aforementioned FCO-IM for ORM (Bakema et al. 2005) versus Terry Halpin's ORM 2, and the UML-like notation in iCOM (Fillottrani et al. 2012). Extensions include, among others, OntoUML for ontology-driven structural conceptual data models (Guizzardi et al. 2018), TREND for temporal conceptual data modelling (Keet and Berman 2017), and MADS for spatio-temporal conceptual modelling (Parent et al. 2006).

[3] The most recent details can be found in (Fillottrani and Keet 2021).

7.1 A Beauty Contest

Some researchers went ahead and defined extensions to DL-based OWL species regardless, which mostly concern uncertainty—fuzzy, probabilistic, rough—and a few temporal OWL flavours.[4] With Common Logic one can get the most precision of the three, having the expressiveness of full first order predicate logic. But freedom to modify is absent because it's standardised as well. Whether modifications and extensions are needed—hence, greater freedom—is, without a doubt, debatable.

Finally, the precision with Ontology is as high as needed with as much freedom as you want, including inventing your own language, and no-one will complain. Well, there may be a muttering here or there when the logic is obscure and badly introduced, but that can be fixed with good presentation. What was intriguing about the freedom, was that when implicit language choices of the staple first order predicate logic were spelled out[5], both some ontologist reviewers and paper presentation attendees were voluntarily insisting on wearing eye caps. They had a blind spot when it came to the ontological commitments embedded in that language and not conceiving of the possibility that one might want something else as basis. Even there, people commence with the default and devise another logic only if they can't find workarounds or can't force it into first order logic, rather than first determining what the language should be like to represent best what they want to represent.

Arguably, it's a knack of computer scientists to even think that modelling languages are up for debate, for prodding, poking, and redefining. The amount of conceptual data modelling languages counts well into the double-digits, there are surely over a hundred logics designed in the context of computation, and there are a whopping 8945 programming languages.[6] Only a small subset is widely used. What it does do in effect, is explore the plethora of possibilities and trade-offs of language features, notations, and practicalities, and then something will stick for some reason or other.

7.1.1.3 Software Assistance and Development Methodologies

Comparing the methods and methodologies, the paucity of modelling support for mind mapping, biology models, and ontology comes, perhaps, as a surprise. On reflection, it can be explained, I think. While mind maps and biological models are staple in education, they are given to the learners, not that it's a regular task to design them from scratch. And if you don't need to design them, there's no need to teach a procedure for it. Philosophical inquiry does have its *mores*, which are passed on

---

[4] See (McGuinness and van Harmelen 2004; Motik et al. 2009) for the OWL species. It is easy to make mistakes that accidentally modify OWL Full or OWL 2 Full. Perhaps because of that, or perhaps because of the lack of ample tools or computational complexity, neither OWL Full nor OWL 2 Full have gained widespread adoption. For uncertainty extensions, see (Bobillo and Straccia 2011; Lukasiewicz and Straccia 2008).

[5] Described in (Fillottrani and Keet 2020) and presented by me at the conference.

[6] They are conservative estimates. The exact number of programming languages is according to the Online Historical Encyclopaedia of Programming Languages, which can be accessed at https://hopl.info/; last accessed on 18 April 2022.

embedded in the pedagogy rather than cookbook style. I suspect that if they were to attract the numbers of students that we have in computing, with the corresponding lecturer:student ratio, then written instructions and automated marking of adherence thereto would emerge as a means to scale up instruction.

It's harder to justify the values for software support listed in Table 7.1. To make hard claims based on facts, it needs an audit of tools, and thus the values given are only my impression on it. Modelling tools do seem to fit a regular pattern. There's a larger number of tools for more widely used types of models and for relatively inert and standardised languages than for languages with a smaller user base and not standardised languages. That emanates from a cost-benefit analysis: the investment in the software development pays off faster with a larger user base and fewer software updates. Also, language stability allows for the available resources to be used on fixing software errors, resulting in more stable tools, and end users have lower tolerance for less-than-convenient application installation instructions than software engineers and researchers.

Abundance and quality of software tools for modelling, however, do not imply tooling for runtime usage of the models themselves. Mind maps and biological models are hardly, if ever, used during runtime in applications; the model is the end product. For especially ontologies, the model is *not* the end product, but a means to an end. The end product is the ontology-driven information system. The paucity of robust infrastructure for developing an ontology is in stark contrast to the relative abundance of their use. Conceptual modelling tools occupy a spot in between the two: there are several reliable modelling tools and a few runtime applications that deploy them. Most tools still assume the conceptual model to be an end product rather than a means to an end; if it's intended as a means beyond database schema creation or code generation, one may be able to exploit OWL. Put differently, there are opportunities for additional software functionality for all types of models.

In sum, as to the beauty contest, be it model types or their languages, the answer to 'mirror, mirror on the wall' is that none of them is the fairest of them all. If you need a model for a first pass structuring of matter, the rigour and expressiveness of ontology and ontologies is a hindrance. If the model needs to be used at runtime for automated reasoning, it must have a formal specification, making sketches and playful pictures unsuitable. And so on for each model type. It may still be feasible to determine which of the flavours are the tastiest within a type of model. For instance, why EER is better than UML class diagrams for designing a database, or why OWL beats OBO to find contradictions. But for the scope of this book, rather than squabbling about the minutiae and in-fights, the focus is to determine which model type is the most relevant to what you, dear reader, would like to accomplish. The presented comparison will assist with that in any case.

7.1.2 Comparison by Example

Besides theoretical arguments for comparing the types of models, we can do this concretely by way of example. In theory, examples illustrate the theory, test the

7.1 A Beauty Contest 149

theory, or serve as a basis from which to develop the theory. There's a danger of cherry-picking examples and extrapolating too much from the insights obtained, in the process missing features that would come to light with other use cases. Mindful of the example-based approach, let's do it anyway. We've revisited the subject domain of dance throughout the chapters and each model looked unlike the other, which is unsurprising given that they are examples of different types of models. We shall compare them briefly first. Afterwards, we'll take a task-based approach: learning content from a textbook through modelling as a method of learning.

7.1.2.1 Comparing the Models About Dance

We have developed five models of dance, from the brainstorming with mind mapping to the ontology of dance. Are we any the wiser? Is there a 'best' one? It's like comparing apples to oranges, nay, apples to *vetkoek*, *arancini*, and *booze balls*: they're all round-ish, edible, and tasty, but... Let's try to compare them anyway. Looking back at the mind map about dance in Fig. 2.3 on page 20, it looks woefully disorganised with content that is all over the place: a half-baked attempt at a taxonomy in the top-right corner in the 'Latin' branch, examples in the 'Ballet' branch, and a hodgepodge under 'Ballroom' with types of dance, a movie title, and a related activity. But it is colourful and has a good layout, especially compared to the lyrebird dance show in Fig. 3.7 on page 44. That figure, whilst focussed on one topic, looks busy after glancing over a mind map and it takes time to digest all the information present in it. It makes the EER diagram in Fig. 4.12 (page 75) look well-structured and less daunting. Comparing these three diagrams to the notation of the salsa ontology in Fig. 5.7 (page 108) and sketch of the ontology of dance in Fig. 6.5 (page 135) is unfair, because the latter is only a sketch for the logic and in ER notation at that and Fig. 5.7 doesn't even have a diagrammatic notation. But if Fig. 5.7 had, it would be either abusing one of the graphical conceptual data modelling notations or an untested ad hoc notation.

But we can compare them as model, that is, their content rather than their presentation. Besides observing that they are small except for the salsa ontology and they exhibit their expected increase in precision, it offers a noteworthy observation about the ontology of dance. The ontology contains generic principles that we saw examples of in the other models. **Organising principle** relates to the Latin and ballroom and the four types of lyrebird dances, **Choreography** to partners, octets and quartets, **Rhythm** to the on-1 and on-2, **Body movement** to body movement, and **Music** to the multimedia audio and sound annotation for the lyrebird shows. Put differently: the dissimilar models actually do overlap in content. With a more comprehensive dance ontology, we could use the ontology to link them all, fulfilling the task that ontologies were intended for.

With overlap in content of the models, there may be opportunities for cross-pollination or cross-purpose use. We could use the mind map for video annotation tasks of dance in general rather than only for salsa videos, albeit less for salsa specifically. We could use the ER diagram of the ontology of dance for ontology development to enhance the structure of the simple salsa ontology and for database development, albeit not for a database for lyrebird data but for recoding information about dances. That database could be populated by all those moves and actions from the salsa ontology. The conceptual data model for the lyrebird database is not easily repurposable; rather, the salsa ontology could be merged into it in the annotation section of it. The biological model about the lyrebird dance sequences is not directly reusable, but with some rework it could be added to that hypothetical chimera of the ontology of dance model and the salsa ontology, in the part on choreography and organising principles.

7.1.2.2 A Task-Based Comparison: Learning About Migrant Labour

Of all the tasks I could have chosen to compare the types of models: why the task of learning about a topic? It's one of the two main uses of mind maps and my original claim that we can do so much more beyond mind maps hasn't been shown when it comes to the same task. So, we need to test whether we can get more out of any of the other types of model than one can get out of a mind map. We should. Moreover, there are two precedents that suggest we must. Just last year, art historian Laura Bertens, of Leiden University in the Netherlands, was looking for ways of visual representations of her course content and in so doing came across the notion of ontologies. Terminologies and ontologies are hard to avoid in arts, with AI for cultural heritage being a well-established niche in academia for at least the past 15–20 years and a range of tools have trickled down into everyday use in the meantime, such as mobile phone-based tour guides that adapt to the type of museum visitor. She wanted her students to think harder about the art history canon, rather than only behave like sponges sucking up received wisdom uncritically, and also to be able to formulate better *how* exactly diverse styles relate rather than only *that* they do, like Fauvism having been influenced by Neo-impressionism whilst another takes stylistic elements from yet another -ism style, and yet others react against a prevailing predecessor style. Practically, Bertens made her students install the Protégé ontology development environment and start modelling the art history canon to clarify the famous flow diagram by Alfred H. Barr that has only names and unlabelled arrows. It made the students uncover hidden assumptions, formulate questions, and dig deeper in an active learning process. According to Bertens, it's mainly the process of modelling that is valuable and perhaps not so much the resultant models themselves.[7]

The other precedent is the textbook complexity assessment by Peter Bollen from Maastricht University in the Netherlands. In his role of lecturer, he wondered how to assess whether a textbook is at the right level for his students. Conversely, if you are or were a student or learner, you may ask yourself how you do, or did, assess whether the textbook assigned for the course was 'easy', 'at the right level', or 'too difficult' and likewise for the amount of course content. The level of the language used to explain the matter may contribute, like short or long sentences, but that also depends on the subject domain and may not hold across disciplines. A key parameter is about how dense the text is with new content. How many new

---

[7] (Bertens 2022).

terms and concepts are introduced and how many relations and constraints between them are described? And precisely that can be captured in conceptual data models, or so went Bollen's reasoning. He experimented with an operations management and a marketing textbook. The set-up was to create an ORM conceptual data model per page for a random selection of textbook pages, and then compare the models for each of the textbooks on their complexity regarding a set of metrics, such as how many entity types, fact types, and constraints are included in the models and how many example instances are given. The differences showed up in the diagrams indeed.[8] And if it shows up in his experiment, and it is reproducible, then we should be able to use the 'modelling a textbook page' approach to learning a page of text as well.

Let's put it to the test. For textbook, I selected one of the two UCT Open Textbook Award winners of 2022, which is on migration in Southern Africa. One of the two editors, Nomkhosi Xulu-Gama, who is also at the University of Cape Town, is the main author of a chapter that zooms in on worker education of foreign nationals. An extract of the begin of its section 5 is included in the text box below.[9]

---

5 The Labour Migration Policy Framework

Among other things that the Employment Services Act No. 4 of 2014 seeks to achieve is the promotion of employment of young work seekers and other vulnerable persons. The act also aims to facilitate the employment of foreign-national migrants in a manner that is consistent with this act and the Immigration Act No. 13 of 2002. According to South African policy and its legislative framework, a 'foreign national' is an individual who is not a South African citizen nor has a permanent residence permit issued in terms of the Immigration Act. The Employment Services Act is designed to be a short-term measure to bridge the skills shortage within the employer's business. It makes clear that foreign-national migrants should be employed on the basis that their employment promotes the training of South African citizens and permanent residents. This consequently excludes many semi-skilled and unskilled migrants from participating in the South African post-apartheid workplace. Additionally, the act states that employment of the foreign national migrant must not impact negatively on existing labour standards or the rights and expectations of South African workers, and it gives effect to the right to fair labour practices as explained by section 23(1) of the Constitution of the Republic of South Africa, Act 108 of 1996 (Republic of South Africa, 1996).

(continued)

---

[8] (Bollen 2006).
[9] (Xulu-Gama et al. 2022).

> Leliveld (1997) documents the negative effects of the restrictive South African labour policy as a shift from the 1994 democratic era:
> Section 1 of the Immigration Amendment Act No. 13 of 2011 defines work as:
>
> > conducting any activity normally associated with the running of a specific business or being employed or conducting activities consistent with being employed or consistent with the profession of the person, with or without remuneration or reward. (Immigration Amendment Act No. 13 of 2011: section 1).
>
> Section 1 also says that a foreigner is an individual who is not a citizen. An illegal foreigner refers to an individual who is in the Republic in contravention of the act.

To start with the status quo in learning, mind maps, the central concept is clearly about **Labour migration policy**. From there, there are a myriad of related terms that can be added largely in sequence of reading the passage. Among others, there are the types of workers, differences about foreigners, and a definition of what a worker is. My sample mind map is depicted in Fig. 7.1, which is on the flatter side of the average of how mind maps typically look like. There is a branch awkwardly called **exclusion**, because we can't cross-out a branch about something worth mentioning to be not applicable. Other than that, it is as before: some branches indicate subclasses or specialisation whereas other are examples. There's no avoiding the overloading of meaning. Yet, mind mapping is evidently a feasible possibility for the task.

Next, taking the biological models approach for the workers doesn't work. The text describes mainly the 'what' components rather than the 'how' processes that biological models are better at. To make matters worse, trying to devise pictograms for an **Employment** concept or the **Fair labour practices** aggregate is far from trivial, if they can be drawn at all.

A conceptual data model can be drawn, as expected from the research. In line with the focus of Chap. 4, I've drawn one in EER notation rather than Bollen's ORM,

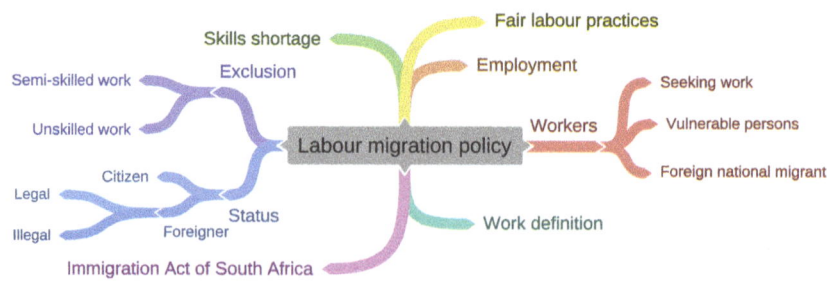

Fig. 7.1 A mind map of the immigrant labour textbook page

7.1 A Beauty Contest 153

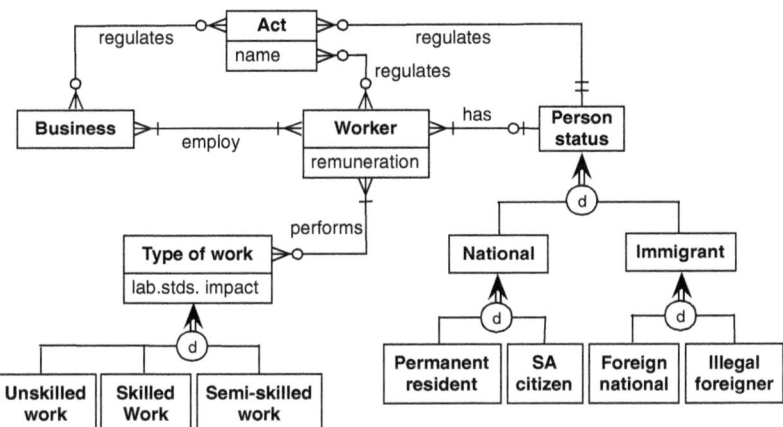

Fig. 7.2 Preliminary EER diagram of the immigrant labour textbook page; the remuneration attribute may be realised as a Boolean yes/no and the labour standards impact (lab.stds. impact) could be a selection from {positive, neutral, negative}

which is included in Fig. 7.2. The model doesn't cover everything of the selected section on migration. Like the mind map, the EER diagram isn't good at dealing with negation, on the exclusion of unskilled and semi-skilled work, and there is simply too little information in the text to merit inclusion of the fair labour practices. On the bright side, now there is a differentiation for the subclasses and explicit relationships with constraints holding over them. Or: for what we have in the model, it is more precise than the mind map. Thus, conceptual data modelling may be as good a choice for modelling the information to complete the task as mind maps are, if not better.

Ontologies? They are an option, too, in agreement with Bertens' results from the art history class. Compared to the conceptual data model, the minimum remodelling required includes elevating the attributes to classes and relationships, adding details on the types of work, and reworking the messy taxonomy of the Person Status. Permanent residents immigrated once upon a time and still possess a nationality other than South African. Also, in other countries, terminology is different and foreigners are carved up variously, such as the three categories of 'aliens' in the USA, which refer to humans from outside the USA in the context of immigration (excluding the non-humans from outer space in the movies) and the European Union's 'long term resident' that doesn't have the same ring to is as South Africa's 'permanent resident', even though their definitions suggest they refer to the same thing. This drags in additional knowledge compared to what was written on the sample page from the textbook. That is good for students exploring the topic more comprehensively, but it may end up being well beyond what is needed for the task. To compare it graphically nonetheless, I created a version "0.1" of the ontology—under the assumption that a future v1.0 counts as a release—about content from the page and without pulling in too much external knowledge, which is shown diagrammatically in Fig. 7.3. Overall, I deem developing an ontology more

154   7 Fit For Purpose

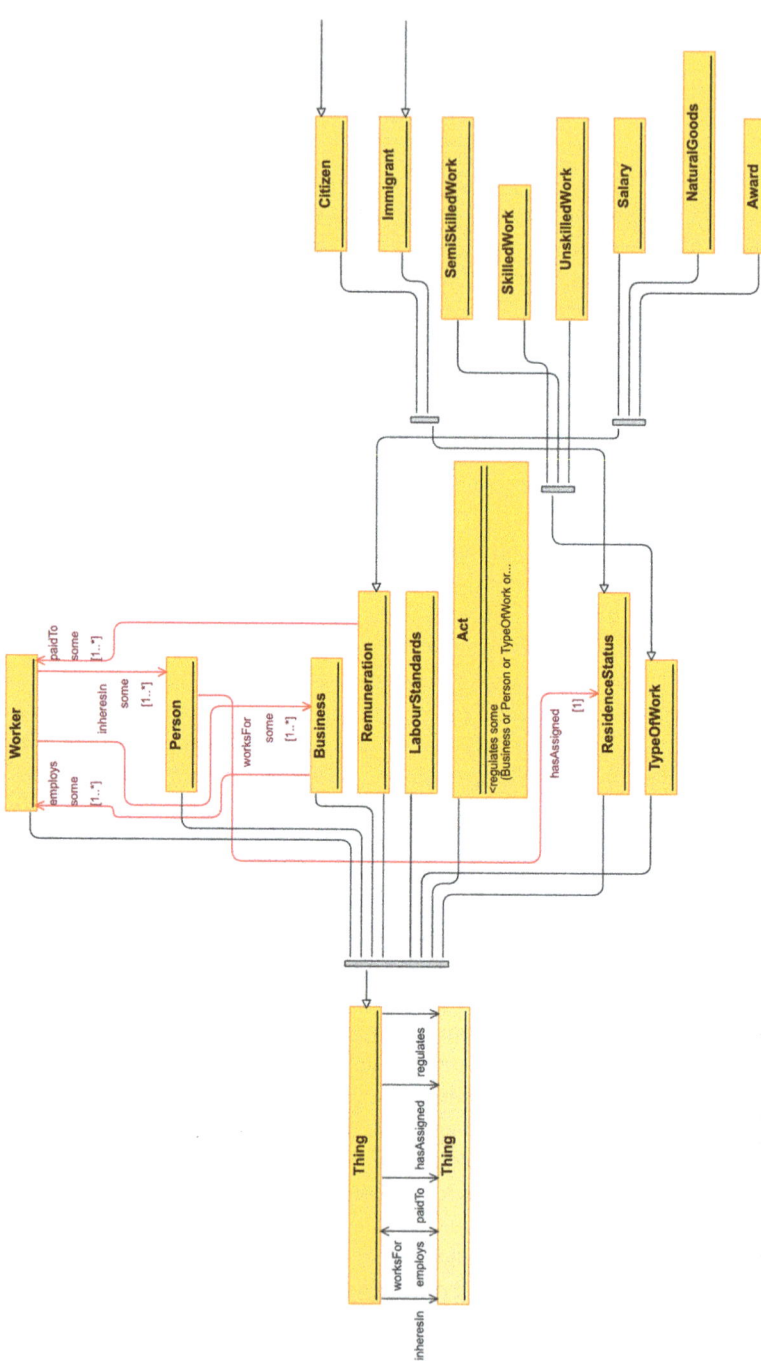

Fig. 7.3 Diagrammatic rendering of a section of the draft version 0.1 ontology related to the immigrant labour textbook page

appropriate for an advanced learning exercise on the theme than for representing a page of text. In education terminology, it's a learning activity at levels 4–5 (analyse and synthesise) of Bloom's taxonomy, rather than the 1–2 (reproduce and comprehend) where mind maps are.

Finally, Ontology. Fine topics to tackle are interrogating that definition of work and where the boundaries lie between unskilled, semi-skilled, and skilled work. In the context of the various acts and worker education of foreign nationals, they are tangential topics, however, and their investigation is unlikely to be within the scope of an undergraduate course on theories, laws, and practices of immigration. Put bluntly, an ontological investigation is over the top for the task at hand.

In sum, for the task of modelling for learning when restricted to just grasping the provided content within sociology, mind maps and conceptual data models seem the most suitable options, and for further engagement, an ontology.

7.2 Ethics and Modelling

Modelling and fit for purpose may not seem immediately related to ethics, but it does in two ways. The first one is the straightforward engineering and technology sense of fit for purpose and a professional's responsibilities toward colleagues and clients. The second one is on bias, which is often lamented about with large language models and deep learning-based models and so we need to consider it. If bias creeps into the types of models we've covered in this book, then we need to know where it may do so. We'll consider each in turn.

7.2.1 Professional Behaviour and Practices

Let me start with a general example on fit for purpose in the context of engineering. Say you have to put a long nail into a piece of hardwood. We know very well that trying to do that with a pool noodle won't work, and so it would be unethical to try to sell pool noodles as hammering devices. Conversely, advertising a hammer as a pool toy to fling around to other swimmers is definitely not good advice either and just as unethical to promote it, be it by a professional or anyone else. Misuse and advice toward misuse is abuse of power bestowed on a professional. A non-expert needs to be able to trust the advice of the expert to be sound advice.

More precisely, engineers and technicians as members of engineering societies, and similarly for specialisations in computing and IT that includes the modellers, have codes of conduct, codes of practice, and/or codes of ethics. For computing, there's the international-but-USA-centric Institute of Electrical and Electronics Engineers (IEEE) and the Association for Computing Machinery (ACM), and there are multiple national organisations, such as the British Computer Society/The Chartered Institute for IT in the UK and the Institute of IT Professionals South Africa. Such codes have imperatives like "Be honest and trustworthy" in a code of ethics and "Give comprehensive and thorough evaluations of computer systems and their impacts, including analysis of possible risks" in a code of good practice.[10]

Applying such imperatives to the engineering sense of fit for purpose, it may be argued that it's wrong to suggest using mind mapping to develop a relational database or object-oriented software. This because mind mapping is for sure not fit for that purpose, most notably because there are no constraints in mind maps, and it would then result in databases that may store data that contradicts the given specifications for the universe of discourse. Conversely, if a group of businesspeople were to go through the laborious process of ontology development for maintenance of the cell phone towers' energy supplies when they only need to brainstorm about how to fix the pressing shortfalls in electricity so as to not have them break down in two days' time, that were to be irresponsible. Then quickly drawing a mind map would be the modelling strategy as fit for purpose. It may still serve the long term to develop an ontology about energy supplies and cell tower maintenance, and it surely would be of use in several countries, but that is then part of another programme of work as compared to the "war room" of the cell phone company to stave off imminent disaster during a particular bout of prolonged nation-wide rolling blackouts. This scenario of the energy supply shortage is real in South Africa; I added the modelling fit-for-purpose to it.[11]

Bear in mind a type of model's purpose, use each one accordingly, or have good, explicit reasons to deviate from it. In most cases, breaking the rules is not a virtue and if you see doing it yourself with these model types, it's more likely the case that either you chose the wrong type or the goalposts of the task shifted and you need to move with it.

7.2.2 A Model's Features and Modelling Pitfalls

Ethics is relevant to the content of the models and what you do with them. The key question is: do those issues of 'ethics in AI' that keep hitting the news headlines exist also for this category of models? The short answer is: no. Those problems with bias can't creep in in the way it does with those data-driven models. It's one thing to learn undesirable patterns from the data of which you didn't know for sure they were implicit in the data, but it's quite another level to create an ontology where a human encodes that explicitly, alike "All programmers are male" or "If black then guilty". Then it's an act of commission rather than one of omission. That is, what for

---

[10] The BCS's information can be accessed at https://www.bcs.org/ and that of the IITPSA at https://www.iitpsa.org.za/. Source of the quoted imperatives: IITPSA's Code of Ethics of July 2021, https://www.iitpsa.org.za/wp-content/uploads/2021/08/IITPSA-Code-of-Ethics-July-2021.pdf (last accessed on 29-5-2023).

[11] The mention of "war room" was made by Jacqui O'Sullivan, Executive of Corporate Affairs at MTN South Africa, during an interview on SAfm Sunrise on 4 July 2022 when she appealed to small businesses to help out with additional generators and community support to prevent battery theft.

7.2 Ethics and Modelling

the biased data-driven models may be a case of "in hindsight, we could and maybe should have guessed as much", it's an intentional conscious act of prejudice or -ism upfront during the active modelling by humans. One may hope there are sufficient people looking at the model before signing off on it.

But what about the less obvious prejudices and biases? Are there guidelines to prevent them or checklists to verify against? If there were guidelines, it would have been included in each of the 'how to' sections in the previous chapters. Ethics mainstreaming, if you will, instead of an afterthought. We're at the afterthought stage, but mostly it's a blank slate on ethical issues in conceptual models. Could it simply not be a problem? Have any issues come up that can be blamed on a conceptual model? The concrete examples from the scientific literature are, to the best of my knowledge, less than a handful; a few others may happen in the future, i.e., the potential for damage. No 'bad boy' headlines for our models is a good thing. As to the sporadic concrete occurrences, their underlying modelling causes fall in either of two categories: manipulating the properties of a class and aggregating too much to hide unpleasant facts. I'll illustrate each one.

7.2.2.1 Property Manipulations

There are two ways to actively manipulate properties for ulterior motives: to either include them or not, and, when included in the model, to manipulate their range by tweaking the kind of objects that may participate or the value range they're permitted to have. The consequences are that either more or fewer objects are classified to be an instance of that class. Imagine you're a player in the trinity of medicine, pharmaceutical company, and health insurer, where the benefit for one party may be a loss for another. With millions, if not billions, of dollars at stake, it may be tempting to fiddle a little with a property's values of a class in one of the models that codify knowledge about diseases, symptoms, and treatment plans, such as SNOMED CT and ICD10. Those models are integrated with electronic health record systems that hospitals use as ontology-driven information systems. Doctors classify a patient as having a disease based on the symptoms and the system returns a list of diagnostic tests and a treatment plan that a healthcare insurer will pay for. For the sake of example, let's say the domain experts from medicine and health insurance deem a person to have hypertension when the blood pressure's values are higher than 140/90 but the domain experts from the pharmaceutical industry claim it is so when the values are higher than 120/85. If it's the latter, more people will be classified as having hypertension and be ordered to take blood pressure-lowering medication. That's good for the pharmaceutical industry that will sell a larger number of pills, but not good for the healthcare insurers when they have to cover the costs of those extra pills. Many a domain has properties where the boundaries of its values are argued about, be it because it hasn't been scientifically determined or the there are no crisp cut-off points.

It was a real issue for determining the boundaries of normal hormonal fluctuations during menopause. There are also real case studies for the more or less properties for a disease or disorder. An example investigated and reported in the scientific literature is about when someone is categorised as an alcoholic or deemed to be suffering from alcohol use disorder.[12] Instead of delving into those details, let's use a concrete and easily accessible example with published facts, in a scenario for a website with real estate listings or with home improvement equipment and projects.

Such a website will have a database back-end where the data for each dwelling is stored, which had a preceding conceptual data model for its development. It may also have a model embedded in the front-end to facilitate search, like when a user wants to select desirable features of dwellings, such as number of bedrooms, and whether it has an air conditioner or heating and of what type. They also could be part of the conceptual data model or originally captured in an ontology that is then used for annotation and search. We'll zoom in on the air conditioners. The US guidelines for air conditioners have well-rounded numbers for the imperial system of measurements, like that for a home of 600–1000 square feet, an air conditioner of 18,000 BTU is advised, according to the HVAC company, and a 14,000–18,000 BTU one for a 700–1000 square feet according to a consumer review.[13] Across the pond, in a Europe with the metric system and European Union regulations for saving energy, they measure in m^2 and W or kW; those USA ranges of square feet and BTU amount to 55.7–92.9 m^2 and 4103–5275 Watt.[14] Following one of the EU guidelines, and the model's properties emanating from it that we need to represent in our model: first, it's per cubic metre, not just surface with square metre, since it's the amount of air in the room to replace and a room is 3-dimensional; second, the level of insulation of the building has to be factored in, as well as what's happening in the room, which adds two multipliers to count for heat transfer through walls, windows, or the roof, and whether you're having all your computer equipment in said room.[15] Then, for that 55.7 m^2 and a regular 2.5 m high ceiling that amounts to 139.5 m^3, and for an average warm room, we get to 5570 W. The 92.9 m^2 with a 2.5 m high ceiling and the worst insulation comes to about 11,612 W (or 39,623 BTU) and with a 2 m low ceiling and best insulation, 5574 W (or 19,019 BTU). That's different from the US guidelines. They can't be both right.

The air conditioner guidelines can be assessed experimentally for a set of rooms in various conditions. For medicine and health, to a certain extent, normal ranges and averages for the various molecules in our bodies can be determined as well, by sex, age range, height, weight and so on if necessary. Where ethics comes into play is the *tweaking* of the values for an ulterior motive. It's hard to imagine that they typically would be modelled due to a cognitive bias as a mistake that slips

---

[12] See (Wakefield 2015) and (Keet 2021) and references therein for further information.

[13] https://www.hvac.com/air-conditioners/ac-size-guide/ and https://www.baltimoresun.com/consumer-reviews/sns-bestreviews-home-best-portable-air-conditioner-2021-20210525-mxdkhcyebzacvefon427bsav4q-story.html, respectively (last accessed on 29-5-2023).

[14] Source: https://ec.europa.eu/energy/eepf-labels/label-type/air-conditioners_en; accessed on 13 June 2022.

[15] I selected a language spoken in Europe to be sure it would not include US results, being Dutch: https://www.allesairco.nl/hoeveel-airco-heb-ik-nodig/; accessed on 13 June 2022.

in without realising it at the time. They're conscious choices modellers make, and model users need to be aware of.

7.2.2.2 Aggregation or Granularity

The other tempting candidate to meddle with is granularity or aggregation. The topic is well-known in modelling, albeit not as a mechanism for building bias into a model. It concerns the question of how detailed the model should be and how fine-grained the taxonomy has to be. It's a harmless and practical engineering question and only straddles into an ethical issue if the choice for modelling less detail is an intentional act of omission, as compared to a 'not needed' or a 'ran out of time and will be included in the next version' or a 'pruned because of efficiency'. It may be difficult to determine which case applies, unless there is an annotation that says so. The Gene Ontology did the latter and annotated it accordingly: there is a GO basic where several relations between entities have been deleted compared to the regular version that is the GO, and then there is a GO plus with additional axioms.[16] This is not an ethical issue, but motivated by engineering.

In contrast, conceptual model design for conflict databases is the low-hanging fruit to illustrate bias that have at least a whiff of ethical issues coming off it. Those issues may be intentional or unintentional. Let's say we're tasked with creating a model with a taxonomy of bombing targets. Most modellers are not an expert in this area, but have general knowledge about it from regularly reading the news, so we could have a go at it as first draft to offer to the domain experts in the first client meeting. The taxonomy includes Government building, Residential Area, and Logistics infrastructure. Now, if they are the only categories available in the software system developed from it, then that's what data capturers will record in the database, even if they have more detailed information about the bombings. What is preventing anyone from adding subtypes to Government Building, such as Hospital, State's generic medicine manufacturing plant, Water purification plan, Military base, and Homeland security torture bunker? In an analysis of conflicts, these more fine-grained differentiations may be essential, especially for a side in the conflict that loses hospitals, medicine factories, and clean water as compared to a side in the conflict that gets thwarted in their military and torture operations. System designers or their clients who're on the side of bombing health facilities to smithereens would *not* want the fine-grained hierarchy, because that makes it easier to uncover that they're playing dirty.[17]

Once seen, the potential issue cannot be unseen. A finer-grained categorisation, or an absence thereof, may be intentional or not. It even may change over time when the subject domain changes or when alliances shift. Besides hypothetical examples for illustration, there are real ones. They're tricky to detect and require elaboration to explain, unlike those stories about deep learning model fails. One such application with a problematic model at the back-end computes a "Dirty War Index" (DWI) that

---

[16] http://geneontology.org/docs/download-ontology/ (last accessed on 29-5-2023).
[17] Examples from Keet (2021).

was to serve as a "public health and human rights tool" to examine and monitor armed conflicts to help decide where resources for health care are most deserving to be distributed, as proposed by Madelyn Hsiao-Rei Hicks, a psychiatrist then with King's College London, and Michael Spagat, an economist with the Royal Holloway College, University of London, in the UK.[18] The idea sounds appealing, but the details are less straightforward. The DWI is calculated as ((Number of "dirty" cases)/Total number of cases) × 100, where "dirty" means undesirable or prohibited by law, such as reports on cases of child soldiers, torture, and prohibited weapons. So far so good. It requires a database with the case reports and which party in the conflict is responsible for the recorded act, in order to be able to calculate the DWI for each side in the conflict and determine who's the dirtiest of them all. A challenge to make this work is data collection, as it's known to be difficult to obtain reliable information in conflict zones both regarding who's to blame and what gets reported. That's independent of modelling, however, so we're going to abstract away from data collection issues and assume the data are being recorded adequately. The issue we'll look at are the databases from where to extract the information, which each have a conceptual data model at their inception.

The database Hicks and Spagat use for one of their main illustrations is the fine-grained CAIN web service recording deaths during the protracted conflict in Northern Ireland from 1969 to 2001.[19] They use the "Sutton Index of Deaths" to calculate two complementary DWIs, being "aggressive acts (killing civilians) and endangerment to civilians (by not wearing uniforms)". A first question is the ontology of 'civilian', which is good for a long-standing debate in peace research. What is the definition of civilian and who is one and who isn't? Is an off-duty soldier who's not wearing a uniform a civilian, and a guerrilla general who never has worn a uniform, or the nurse who tends the wounded, or the family who hides a fugitive paramilitary bomber? Less contentious, but not without debate either, is determining what the properties of 'government' are, and what its subtype may be, such as 'legitimate government' and 'occupier'. Currently, there is no standardised model for either. An end-user tool may hide both the questions and the answers its developers or their client had settled on, but their respective meaning affect data collection and analysis.

Regardless, in Hicks and Spagat's view and terminology for the Northern Ireland conflict, the "British Security Forces" (BSF) have a "Civilian mortality DWI" of 52, for the "Irish Republican Paramilitaries" (IRP) that was 36, and the "Loyalist paramilitaries" (LP) had a DWI of 86. Aside from the chosen naming, there are the aggregations of the actors, into three: the official state force, the unofficial state force that does the state's dirty work, and at least one aggrieved group. What does IRP refer to? The Irish Republican Army (IRA) alone, or lumped together with the Real-IRA and Continuity-IRA, and does it include also the political organisation

---

[18] (Hicks and Spagat 2008).

[19] The web service is available at http://cain.ulst.ac.uk/sutton/index.html (last accessed on 29-5-2023).

Sinn Féin? And LP for all UFF, LVF, and so forth? Why not aggregate by IRP versus BSF+LP? Because it would look too bad for the UK? Following the data trail to the source, interestingly, the CAIN web service lists a whopping 29 groups that were in some way involved in the protracted conflict. Put differently: the details are there, but Hicks and Spagat preferred this particular aggregation for their DWI. Their example about the aggregation of participating groups in the protracted conflict in Colombia does nor fare better. Yet, details on such groups helps conflict analysis, especially when there are shifting alliances. An explicitly declared fine-grained model can provide the sought-after transparency, and therewith gain trust from the user irrespective of their view on the selected biases. I raised such issues in 2009, as an example of the, for the time, key novelty that there may be modelling issues in the models that people design for databases and information systems.[20] It is only in recent years, well over 10 years later, that the topic of bias in modelling is gradually receiving attention and further new insights are likely to be obtained yet.

Wanton aggregation when the data is easily accessible or collectable, or the additional features are known to affect the classification, especially if it changes a conclusion or misleads in perception, is an act of commission, not a hidden bias. The examples in this section are easy to spot once a modeller has enough knowledge of the subject domain and knows of the potential pitfalls. While the former, domain knowledge, is not within my circle of influence, hereby you now know of the latter, aggregation/detail, and have the foreknowledge to avert the problem.

7.3 Design Your Own Modelling Language

Now that we have taken a stroll along a selection of different types of models, seen their pros and cons and an ethical dimension to it as well: which one of them do you prefer most? And within that type of model, which one (where options are available)? If your answer is any one of 'neither', 'none', 'not sure', or 'can't make up my mind': what will your ideal modelling language look like if you had the chance to design one yourself? It is doable to answer this question, regardless of whether or not you're a novice in modelling or modelling languages design. Making up your own language, be it for fun or for a specific task or broad category of tasks, can be cast as a design process. There are procedures for that that can be followed—like baking a cake according to the recipe, but then a 'how to design a language' cookbook instruction. Think about it: all those languages did not fall out of the sky. Someone, or, commonly, a group of people, designed them, too. Most modelling languages were not developed in a systematic way, but also modelling is growing up as a field of specialisation. And some modelling language developers did use a procedure or part thereof. The development of OWL went through a requirements specification phase, goals were formulated, the language was designed, and it

---

[20] I discuss mainly the reliability of the DWI tool and tools in general to assist in emergency support in war-torn countries in (Keet 2009), generalising from that only near the end of the paper.

was assessed on usage for a 'lessons learned' phase, which fed into a round of improvements.[21] The Distributed Ontology, Model, and Specification Language development process also went through the requirements phase and its developers and user base extracted modelling use cases from existing reported issues to inform the designers what the language should be able to express to solve those problems. One of the modelling tasks was about the mereology we've seen before.[22]

Methods and methodologies for designing languages have been proposed. They are not widely used yet, but even so, the design approach to a modelling language casts a fresh light on both the languages you have come across and the notion of fit-for-purpose. And it may be just fun to think outside the box: besides sponging the received languages, you can have agency just like the designers of those languages. I can't promise your newly designed pet language will gain traction, because there are many factors that determine success, but you will be able to boast that you designed your own one.

The procedure that we'll use here was inspired by Ulrich Frank's waterfall methodology for domain-specific languages that my colleague Pablo Fillottrani and I modified to make it work for ontology languages. We didn't set out do so intentionally, but in trying to design evidence-based and logic-based conceptual data modelling language profiles, it felt like we couldn't do without a proper procedure to justify the design. Since there was no such procedure for language design for information and knowledge representation, we ended up investigating it as a necessary detour before returning to the original task at hand. We then adapted it further to also be useful for conceptual data modelling languages. The easily observable changes are that we made Frank's waterfall iterative and some of the steps optional; the challenging difference is the step on ontologically fundamental design decisions. Ontological commitments are embedded in each language, even if you thought not.[23] Grosso modo, they follow the same path, which is depicted in Fig. 7.4. Let's look at each step.

Step 1 The first step is about the clarification of scope, purpose, expected benefits, any long-term perspective on the matter, and the feasibility of designing a language in the light of the resources available. For instance, we may want to design a new conceptual data modelling language tailored to model-based data access, that is temporal, and will surpass UML class diagrams. That's unlikely going to work for two reasons: the Object Management Group has extensive resources both in the

---

[21] The process is described in the "the making of an ontology language" (Horrocks et al. 2003) and a digest of the use and rationale for improvements are described in (Horrocks et al. 2006) for the underlying logic and in (Cuenca Grau et al. 2008) for the OWL 2 standard.

[22] The issue with mereology and mereotopology was originally described in (Keet et al. 2012) and the solution with DOL was presented in (Keet and Kutz 2017). Additional use cases for DOL are described by Lange et al. (2012). The DOL is specified in (DOL 2016) and additional technical details are described in (Mossakowski et al. 2015).

[23] They are described in (Frank 2013), (Fillottrani and Keet 2020), and (Fillottrani and Keet 2021), respectively.

7.3 Design Your Own Modelling Language

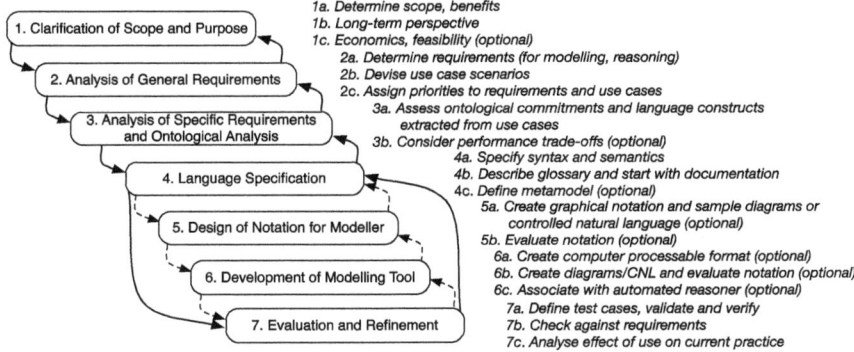

Fig. 7.4 Summary of design steps to design your own modelling language. They vary a little by the type of the modelling language to develop. (Adapted from Fillottrani and Keet 2020)

short and in the long term to sustain and market UML than we have, and temporal constructs are computationally expensive and thus won't scale to lots of data. So, we're halted in our tracks already. That's a good thing, to not waste resources on such an endeavour.

Let's try again. A new temporal conceptual data modelling language that extends UML class diagrams and has a logic-based reconstruction for precision. Its purpose would be to model the subject domain more precisely. The expected benefits are better-quality models and therewith better quality applications. The long-term perspective is not applicable, since it's just a use case scenario. Regarding feasibility, we have, by now, the competencies to develop it and the funding for ourselves to do the design or to do it on a volunteer basis. Funding for software developers to create a tool for it would be good to have, as well as assistants to help evaluating the language and the tool. For the sake of example, let's pretend we have all that.

Step 2 The analysis and general requirements can be split up into three components, of which the first two can be done in any order or simultaneously: determine the requirements for modelling (and possibly automated reasoning as well), devise use case scenarios, and assign priorities. A requirement could be that we need to be able to represent that some entity evolved from being an instance of one class into another one, like that an employee who's a Product Manager may change their role to that of Area Manager, or to be able to check that only those people who have paid their ticket can check in for their flight. Conjuring up a list of requirements out of the blue is not easy, but may be assisted by a library of requirements to select from. For conceptual modelling languages, there is no such library, nor was there one for ontology languages when we needed it. So, we created a preliminary library of features for ontology language requirements.[24]

---

[24] Those sample requirements as a 'library' are listed in a blog post, at https://keet.wordpress.com/2020/04/10/a-draft-requirements-catalogue-for-ontology-languages/ (last accessed on 29-5-2023).

Use cases can be varied, depending on the scope and purpose. For determining requirements and, eventually, meeting them, use cases can be described as the kind of things you want to be able to represent in the prospective language. For instance, if a product manager may change into an area manager, we need a language feature called object migration, and for representing that an employee can be assigned on a project for a specified duration, a ternary relationship, although a stakeholder specifying these examples may not know that. Another type of use case may be about how a modeller would interact with the language and the prospective tool.

Step 3 This is the step about ontological commitments, where insights from Chap. 6 and from other philosophers can be useful, if you care. With our prospective temporal UML version, we implicitly adhere to the most common take on representing temporality, where there are 3-dimensional objects and a separate temporal dimension is added whenever the class needs it. The other option is 4-dimensionalism, where time is embedded in all entities, as if they're space-time worms. In philosophy, the former is called endurantism and the latter perdurantism. Different views affect how time is included in the language. Another example is that the main, if not all, conceptual data modelling languages are positionalist on the ontology of relations. It means that there's a relation and there are participants playing roles in it, where roles are components of the relation. For instance, when John loves Mary, John plays the [lover] role and Mary the [beloved] role in a relation loving, and as long as we know those roles, we can jumble up the order of the entities in the relation and in the sentence constructions, like 'Mary is loved by John' and 'it is Mary who John loves'. The three sentences are simply different linguistic realisations of the same relation between John and Mary, the same underlying state of affairs in reality. It does not admit 'Mary loves John', because the sentence assigns each participant a different role. If you conflate natural language and semantics of the logic and reality—called the standard view—then 'loves' and 'loved by' would be two separate relations. This is the case in plain vanilla first order predicate logic, most Description Logics, and OWL. It seems like the logicians don't like the extra machinery in the logic. Philosophers have been trying to characterise relations also differently, without positions and no ordering of participants, but the jury is still out.[25] Ontological differences as to what relations really are, they are indeed.

There are more such ontological decisions, besides these obvious ones on time and relations.[26] Until recently, they've mostly been left implicit, and time will tell whether it'll be picked up for the design of new languages. The performance trade-offs sub-step in step 3 always has received a prominent place if logic plays a role. Which elements are going to be in the language, how are they going to look like,

---

[25] Three ontologically distinct notions of relations, and that terminology of standard view, positionalist, and anti-positionalist were first extensively described by Fine (2000) and later elaborated on mainly in Joop Leo's paper leading up to his PhD thesis on the topic (Leo 2008).

[26] They are described in (Fillottrani and Keet 2020).

7.3 Design Your Own Modelling Language

how scalable does it have to be, and should it extend existing infrastructure or be something entirely separate from it? To answer this for our temporal UML: the atemporal elements will be those from UML and the temporal stuff shall be carried over from the TREND conceptual data modelling language that's for EER, and the underlying logic, $\mathcal{DLR}_{\mathcal{US}}$, is not even remotely close to being scalable. Of course, it's possible to make other decisions here.

Step 4 In step 4 we're finally getting down to the business of defining the language. There are two key ways of doing so: either define the syntax and the semantics or make a metamodel for the language. The syntax can be alike we've seen in Fig. 4.2 for EER: the graphical elements you can use, and how you can use them. This we can do also for logics, like that UML's arrow for class subsumption is a $\Rightarrow$ in our logic-based reconstruction, or $\rightarrow$, or $\sqsubseteq$, or as you wish, or we can modify the UML arrow to our liking.

This first aspect settled on, we now have to give it meaning, or: define the semantics of the language. For instance, that a rectangle means that it's a class that can have instances, and that that fancy arrow means that if $C \Rightarrow D$, then all instances of C are also instances of D in all possible worlds; that is, that in the interpretation of $C \Rightarrow D$ we have that $C^I \subseteq D^I$. Logic may be unfamiliar and anyhow one may prefer diagramming, for which metamodelling to define the language may be a way out. It gets many students confused when we ask them to design an EER diagram of the EER language. Sometimes a language can be defined in its own language, sometimes not—designers like to do so, and for sure ORM can be.[27] For our temporal UML example, we can use the conversions from EER to UML class diagrams, and then also reuse the extant logic-based reconstruction in the $\mathcal{DLR}_{\mathcal{US}}$ Description Logic.

Once all that has been sorted out and specified, it's time to write the glossary and documentation so that other people can figure out what it's all about. Few people enjoy the activity. The UML standard is over 700 pages long, but, to be fair, it covers also other content besides the UML Class Diagram specification. The Distributed Ontology, Model, and Specification Language is 209 pages, and the Common Logic Standard is 70 pages. Others obfuscate their length by rendering it as a web page, as is the case of OWL and RDF. I copied the OWL 2 functional style syntax in A4-sized MS Word, which amounted to 118 pages in 12-point Times New Roman font. The original syntax and semantics of the underlying logic, including the key algorithms, is just 11 pages or about 20 reformatted in 12-point single-column A4 format.[28] The main differences? Notation, explanation, examples, and figures to

---

[27] This issue is described for ORM in (Halpin 2004). For a recent account on metamodelling for EER, UML class diagrams, and ORM and, notably a unifying metamodel for them, consult Keet and Fillottrani (2015).

[28] The DOL standard of 2018 is available online (DOL 2016), as is CL standard of 2018 (https://www.iso.org/standard/66249.html) (last accessed on 29-5-2023); OWL 2: https://www.w3.org/TR/owl2-syntax/ and https://www.w3.org/TR/2012/REC-owl2-direct-semantics-20121211/, respectively. The 11-page $\mathcal{SROIQ}$ paper: (Horrocks et al. 2006).

make it accessible to a broader readership. Honestly, I'd suggest you only do that when you're happy with your language and passed steps 5–7 as well, since having to update the documentation is less pleasant than writing it. The prospective temporal UML specification will take up a large number of pages.

Step 5 Design the notation. One could argue that the language specification is *the* notation, but, practically, different stakeholders want different things out of it, especially if your language is more like a programming language or a logic rather than diagrammatic. Depending on your intended audience, graphical or textual notations may be preferred. You'll need to tweak that additional fancy-looking notation and test it with a representative selection of intended users on whether the models are easy to understand and to create. That never happened at the bedrock of any of the popular logics, be it first order predicate logic, Description Logics, or OWL, which may well be a reason why there are so many research papers on providing a nicer version of them, sugar-coating it either diagrammatically, with text based on a controlled natural language, or another syntax. OWL 2 has five official syntaxes and many unofficial diagrammatic and natural language renderings. For our temporal UML: since we're transferring TREND, we may as well copy over the tested graphical notation and adjust the existing controlled natural language to UML terminology.

Step 6 Create a computer-processable format of it, also called a serialisation. This assumes you want to have it implemented and a modelling tool for it. This need not be a requirement for your toy language and you could just as well skip this step. Creating such a format (if the ones of steps 4 and 5 weren't designed for it) will help getting it adopted beyond yourself. It's by no means a guarantee that it will, though. There are also other reasons why you may want to create a computer processable version for it, such as sending it to an automated reasoner or automatically checking that a model adheres to the language specifications and highlight syntax errors, or any of the other application scenarios mentioned in this book. Since it our temporal UML is fictitious, it won't have a computer-processable format and neither does TREND to copy it from, although we ought to do it because we want a tool for both modelling languages.

Step 7 Evaluation involves defining and executing test cases to validate and verify the language. Remember those use cases from step 2 and the ontological requirements of step 3? They count as test cases: can that be modelled in the new language and does it have the selected features? If so, good; if not, you better have a good reason for why not. If you don't have a good reason, then you'll need to return to step 4 to improve the language. For our temporal UML, we're all sorted, as both the object and relation migration constraints can be represented, as well as ternaries.

Let's optimistically assume it all went well, and your language passes all those tests. The last task for the first round is to analyse the effect of usage in practice. Do users use the language in the way it was intended to be used? Are they under-using some language features and discovering they want another, now that they're deploying it? Are there unexpected user groups with additional requirements that may be strategically savvy to satisfy? If the answers are a resounding 'no' to the second and third question in particular, you may rest on your laurels for a little while. If the answer is 'yes', it may be time to cycle through the procedure again to incorporate updates and meet moving goalposts. There's no shame in that. There's UML 1.0 of 1997 and then 1.1, 1.3, 1.4, 1.5, 2.0, 2.1, 2.1.1, 2.1.2, 2.2, 2.3, 2.4.1, 2.5, and finally 2.5.1 of 2017. There is a "UML 2.6 Revision Task Force" that faces an issue tracker list nearing 800 issues by now, 5 years after its official release.[29] They are not all issues with the UML class diagram language, but it does indicate things change. OWL had a first version in 2004 and then a revised one in 2008. ER evolved into EER; ORM into ORM 2.

Either way, regardless of whether your pet language is used by anyone other than yourself, it's fun designing one, even if for the sole reason that you don't have to abide by someone else's diktat of what they decided the modelling language should be. If it's about the same as an existing language, then now you'll understand the rationale of its developers. What the procedure doesn't include, but is also fun and may help marketing your language, is how to name it. UML, ER, and ORM are, to be honest, boring acronyms and not easy to pronounce. Mind Map is a fine alliteration. OWL for the Web Ontology Language is nifty in that an owl as animal is associated with knowledge, and OWL is a knowledge representation language. That certainly beats WOL, albeit being a tad bit long for explanation. Temporal ER languages sometimes have good names too, like TimER and TREND.

And with that fanciful cherry topping of naming a newly created language, we have pushed it as far as possible.

References

(2016) Distributed ontology, model, and specification language. Tech. rep. http://www.omg.org/spec/DOL/

Bakema G, Zwart JP, van der Lek H (2005) Volledig Communicatiegeoriënteerde Informatiemodellering FCO-IM. Academic Service

Bertens LMF (2022) Modeling the art historical canon. Arts Humanit Higher Educ 21(3):240–262

Bobillo F, Straccia U (2011) Fuzzy ontology representation using OWL 2. Int J Approx Reason 52:1073–1094

Bollen P (2006) Using fact-orientation for instructional design. In: Meersman R, Tari Z, Herrero Pea (eds) OTM Workshops 2006—2nd International Workshop on Object-Role Modelling (ORM 2006). LNCS, vol 4278. Springer, Berlin, pp 1231–1241

Cuenca Grau B, Horrocks I, Motik B, Parsia B, Patel-Schneider P, Sattler U (2008) OWL 2: The next step for OWL. J Web Semant Sci Serv Agents the World Wide Web 6(4):309–322

Fillottrani P, Keet CM (2021) Evidence-based lean conceptual data modelling languages. J Comput Sci Technol 21(2):e10

Fillottrani PR, Keet CM (2020) An analysis of commitments in ontology language design. In: Brodaric B, Neuhaus F (eds) 11th International Conference on Formal Ontology in Information Systems 2020 (FOIS'20), vol 330. IOS Press, FAIA, pp 46–60

---

[29] On 30 September 2022, see https://issues.omg.org/issues/lists/uml2-rtf for the latest.

Fillottrani PR, Franconi E, Tessaris S (2012) The ICOM 3.0 intelligent conceptual modelling tool and methodology. Semant Web J 3(3):293–306

Fine K (2000) Neutral relations. Philos Rev 109(1):1–33

Frank U (2013) Domain-specific modeling languages - requirements analysis and design guidelines. In: Reinhartz-Berger I, Sturm A, Clark T, Bettin J, Cohen S (eds) Domain Engineering: Product Lines, Conceptual Models, and Languages. Springer, Berlin, pp 133–157

Guizzardi G, Fonseca CM, Benevides AB, Almeida JPA, Porello D, Sales TP (2018) Endurant types in ontology-driven conceptual modeling: towards OntoUML 2.0. In: Trujillo JC et al (eds) Proc. of ER 2018. LNCS, vol 11157. Springer, Berlin, pp 136–150

Halpin TA (2004) Advanced topics in database research, vol 3, Idea Publishing Group, Hershey PA, USA, chap Comparing Metamodels for ER, ORM and UML Data Models, pp 23–44

Hicks MHR, Spagat M (2008) The dirty war index: a public health and human rights tool for examining and monitoring armed conflict outcomes. PLoS Med 5(12):e243

Horrocks I, Patel-Schneider PF, van Harmelen F (2003) From SHIQ and RDF to OWL: the making of a web ontology language. J Web Semant 1(1):7–26

Horrocks I, Kutz O, Sattler U (2006) The even more irresistible $SROIQ$. Proceedings of KR'06, pp 452–457

Keet C (2021) An exploration into cognitive bias in ontologies. In: Sanfilippo E et al (eds) Proceedings of Cognition And OntologieS (CAOS'21), part of JOWO'21, CEUR-WS, vol 2969, p 17p

Keet CM (2009) Dirty wars, databases, and indices. Peace Conflict Rev 4(1):75–78

Keet CM, Berman S (2017) Determining the preferred representation of temporal constraints in conceptual models. In: Mayr H et al (eds) 36th International Conference on Conceptual Modeling (ER'17). LNCS, vol 10650. Springer, Berlin, pp 437–450

Keet CM, Fillottrani PR (2015) An ontology-driven unifying metamodel of UML Class Diagrams, EER, and ORM2. Data Knowl Eng 98:30–53

Keet CM, Kutz O (2017) Orchestrating a network of mereo(topo)logical theories. In: Proceedings of the Knowledge Capture Conference (K-CAP'17), K-CAP 2017. ACM, New York, pp 11:1–11:8

Keet CM, Fernández-Reyes FC, Morales-González A (2012) Representing mereotopological relations in OWL ontologies with ONTOPARTS. In: Simperl E et al (eds) Proceedings of the 9th Extended Semantic Web Conference (ESWC'12). LNCS, vol 7295. Springer, Berlin, pp 240–254

Lange C, Mossakowski T, Kutz O, Galinski C, Grüninger M, Couto Vale D (2012) The distributed ontology language (DOL): use cases, syntax, and extensibility. In: Terminology and Knowledge Engineering Conference (TKE'12)

Leo J (2008) Modeling relations. J Philos Logic 37:353–385

Lukasiewicz T, Straccia U (2008) Managing uncertainty and vagueness in description logics for the semantic web. J Web Semant 6(4):291–308

McGuinness DL, van Harmelen F (2004) OWL web ontology language overview. W3C Recommendation. http://www.w3.org/TR/owl-features/

Mossakowski T, Codescu M, Neuhaus F, Kutz O (2015) The Road to Universal Logic–Festschrift for 50th birthday of Jean-Yves Beziau, Volume II, Birkhäuser, chap The distributed ontology, modelling and specification language—DOL. Studies in Universal Logic

Motik B, Patel-Schneider PF, Parsia B (2009) OWL 2 web ontology language structural specification and functional-style syntax. W3c recommendation, W3C. http://www.w3.org/TR/owl2-syntax/

Parent C, Spaccapietra S, Zimányi E (2006) Conceptual modeling for traditional and spatio-temporal applications—the MADS approach. Springer, Verlag

Wakefield JC (2015) Dsm-5 substance use disorder: How conceptual missteps weakened the foundations of the addictive disorders field. Acta Psychiatr Scand 132(5):327–334

Xulu-Gama N, Nhari SR, Malabela M, Mogoru T (2022) Policy implementation challenges for worker education and foreign national migrants. Springer, Cham, pp 91–105

Go Forth and Model

8

> *Do your own thing on your own terms and get what you came here for*
>
> — Oliver James
>
> *If you want to be productive, follow leads and dig. Whether it is for oil, gold or information, it requires action - your action.*
>
> — Andrew Saul

Modelling, of the kind discussed in this book, is not the most popular. They used to be new and shiny in their own peak times, when machine learning models were theory and data-based language models would have been the wildest dreams of a few. But just that other types of models win the popularity contest does not mean that the information and knowledge-based models are useless. They never were useless and there's a rightful place for them where the a priori existence of vast amounts of data won't solve the problem or there is no such data or in too limited amounts from which to learn a good model. Or you simply need to do a different task, like learning course material through structuring the content and summarisation, analyse a subject domain, record ground truth unambiguously, design a database or integrate a few, and the many other tasks we can accomplish with conceptual modelling.

A few articles of text can already be used to create a word cloud and it may suffice to bootstrap a model and draw a candidate mind map or to seed a candidate list of entities. We can do likewise for conceptual models and ontologies, contingent on obtaining a larger amount of text to process. They all still need a 'human in the loop' to finalise the model. Text alone has no referent to the reality—the physical world with its societies and unwritten or only partially written mores—whereas humans do, even if imperfectly. Thus, while the algorithms for bottom-up development of models may improve and therewith reduce the human effort in the process, inherently, the human can't be eliminated from it because those automation-only systems lack the connection to physical world.

That potential fear of human redundancy quenched, or safely set aside, the many uses of the models that have passed the revue should have made clear that they and the tasks they are used for can't be replaced by the likes of ChatGPT that had enthralled the populace in late 2022 and early 2023. The responses by ChatGPT were good fodder for jokes thanks to the fluent nonsense it regularly spat out. It may be successful in convincing a user it solved the problem of answering questions well, but that does not imply it solved the problem of answering questions well. The Large Language Models need the symbolic AI—the kind of models introduced in this book, provided they have a fine formal specification—to do the sense-making, to iron out the nonsense part of the answer so that it's both internally coherent in the answer and, assuming the models are good, has had its checks with reality through that (still indirect) way.[1] Even so, the conceptual models and the modelling of them assists with many other tasks, from facilitating learning to database development to making the implicit explicit to resolve misunderstandings.

We've worked through five distinct approaches to represent information and knowledge in a model. As we have seen, there's no best one despite what some may claim. While one approach clearly can solve certain limitations of another, they all have shortcomings of their own. I must admit that it is tempting to guesstimate that conceptual data models occupy a sweet spot in the middle, notwithstanding that I conduct more research on ontologies than conceptual data models. I am also well aware that ontologies are not for everyone and that they are very much a back-end technology in software applications. They're not meant for the end user to put up with, let alone develop. Conceptual data models at least look more concrete and tangible.

Whether each type of model is the best for their own purpose is a separate question. There's only very limited research on proving fit-for-purpose, in no small part because it is difficult to investigate reliably and in a reproducible manner. I'm not going to answer that here and now, either. But we can answer several other questions about modelling. I had fired off five questions in the Introduction in Chap. 1 and those we can answer now at least to some extent.

First, on what type of models there are, I mentioned four main categories of models, being the physical models, mathematical models, data-driven models, and conceptual models, focussing on the latter in this book. We have zoomed in on mind maps, biological models, conceptual data models, ontologies, and ontology. I selected types of models that did fit together in one coherent arc. Modelling events or processes is generally considered to be orthogonal to the models we have dealt with in this book. We can add that to the models and weave a new story. That is, other types of conceptual models exist, notably domain-specific models, models that are graphs, and event or process diagrams. Perhaps an encyclopaedia of models is in

---

[1] A sampling of research that is already heading in that direction is listed in my blog post on the topic, emanating from attending the Empirical Methods in NLP 2022 (EMNLP'22) conference in December 2022, available at https://keet.wordpress.com/2022/12/14/emnlp22-trip-report-neuro-symbolic-approaches-in-nlp-are-on-the-rise/ (last accessed on 29-5-2023).

order. Together with the range of types of models, it does make one wonder whether modelling should become its own discipline. It has already been argued for by Jordi Cabot, then with the Open University of Catalonia and Antonio Vallecillo, with the University of Malaga, in Spain, roundabout the time when I was three-quarters into writing this book. Their 7-page position article echoes an argument I make in this book regarding reuse of modelling approaches and tools from computing in other disciplines, like that I devised a procedure for creating biological models based on my experiences from modelling in computing. Cabot and Vallecillo make optimistic, bolder claims on how far we'd already be toward edging closer to a new discipline, since other aspects from modelling in software engineering can be exported as well.[2] While there's some way to go still, we're certainly moving in that direction and this book may contribute to it.

A discipline needs some commonalities among its components, among others. Besides providing an abstraction from reality, there's one that concerns the second question from the Introduction, on how to build one. Each type of model has its own peculiarities, but there are recurring themes. If you're conversant in one methodology for one type of model, you can find your way developing a model of another type. There are entities and there are relations, and they need to be connected, which can be done in top-down fashion, bottom-up, or middle-out, and after that core design, there are refinement and prettifying stages, and documentation and maintenance. The myriad of methods and tools for ontology development and quality assurance is an overkill for mind maps, but a mind map will roll out of the process nonetheless. Conversely, drawing those mind maps in sprawling directions of ad hoc associations may help setting a scope for the subject domain just like competency questions are an informal way to demarcate the scope of the ontology in the early stages of an ontology development procedure. Internalising such procedures takes effort and building up experience takes time, but an art it does not have to be relegated to. It used to be an art, but years of research did take place and we know much more about the design processes and the notion of model quality now than the 25 or more years ago when each type of model was proposed. The resulting edifice—thanks to all those methods, techniques, tools, and methodologies—may not be an engineering marvel like the International Space Station or the Burj Khalifa, but strides have been made into science-backed engineering. Ignore those advances at your peril. I've seen it happening, where 'ontology developers' don't go beyond the "Ontology Development 101" document from over 20 years ago and then complain their 'ontology' is not pretty or not doing what they hoped for.[3] They have no right to complain. The discipline/specialisation/field of study/area of modelling forges ahead with or without you.

---

[2] (Cabot and Vallecillo 2022).

[3] That document, with that title, is available at https://protege.stanford.edu/publications/ontology_development/ontology101.pdf (last accessed on 29-5-2023). It was cited 7595 times when I last checked it on 2 January 2023, collecting citations still, from papers published in 2022.

Moving on to the next question from the Introduction, the simple question of what you can do with a model is not as easy to answer, for there are many things that can be done with a conceptual model. They all assist with structuring the domain, increasing understanding of it, and sharpening one's analytical skills. Those modelling languages with a higher expressivity also enable precision, derivation of 'new' (more precisely: implicit) knowledge, and communication, be that human-human, human-computer, or computer-computer communication. There are also purposes and suitability for specific tasks that depend on the type of model. It ranges from brainstorming and learning with mind maps, to modelling for database and software development, to data integration, document search, data access, automatic generation questions and answers, knowledge discovery, and up to philosophical inquiry, among other tasks.

We also saw, and as summarised in Chap. 7, that different types of models have different purposes, with a, to a novice perhaps confusing, certain amount of use-function re-purposing and (ab)use. The main purpose of conceptual data models such as EER diagrams is to capture the characteristics of the data to be stored in the database or software application in an implementation-independent way. Their graphical notation is borrowed in ontology development and documentation. Mind maps can be used for brainstorming and very basic structuring of a topic. It could stop there, which it mostly does when it's used in industry and primary or secondary school. It also can be a first sketch for scoping the theme for the prospective ontology that is to power an ontology-driven information system. Biological models are created to visualise biological knowledge emanating from the research. If that's done with modelling software that has a fixed set of icons, they could be used in bottom-up ontology development in semi-automated ontology learning. Ontologies, in turn, are created to represent the knowledge of a subject domain precisely in an application-independent yet computer-processable way. They can inform conceptual data model development and a taxonomy easily could be reused as a section of a conceptual data model or be converted into a fixed set of possible values a class or attribute may have.

Finally, that last question: Why do all that modelling? Because you can't do without it. Well, I suppose you could since you did so before reading this book. More and better modelling, however, may get you ahead of those who don't, thanks to the focus and more systematic domain analysis to structure content, ideas, and theories. As an added benefit, it will also make communication with a data analyst easier, because you now will be able to be accurate regarding the data needs for a prospective application. And that, in turn, should result in tools that behave better according to what you want them to do.

This overview to declarative models and modelling of information and knowledge together with their how-tos hopefully will have given you new insights into modelling. The stepwise expansion would enhance your 'modelling toolbox' so that the toolbox doesn't contain only a blunt hammer or two, but also new nifty tools for more advanced analysis, whilst observing each modelling type's strengths and weaknesses. And if not, then that you've picked up a few nuggets of information about topics as diverse as cladograms, question generation, plankton, lyrebirds, lemonade, labour migration, and dance.

Reference

Cabot J, Vallecillo A (2022) Modeling should be an independent scientific discipline. Softw Syst Model 22:2101–2107

Index

Symbols
ER_{VT}, 57

A
acronyms, list of, xiii
Aggregation
    association, 61, 120, *See also* granularity
Air conditioner, 158
Aristotle, 4, 117
Art history canon, 150
Attribute, 50
    frozen, 59
Axiom, 85

B
Barker notation, 55
Basic Formal Ontology (BFO), 105, *106*
Bertens, L., 150
BFO, *see* Basic Formal Ontology (BFO)
Bias, 155
Bloom's taxonomy, 155
Bollen, P., 150
Bottom-up approach, 70, 103
Buzan, T., 13

C
Carroll, N., 134
Catley, K., 35
ChatGPT, 2
ChemDraw, 38
Chen, P., 52
Choreography, 134
Cladist, 34
Cladogram, 34, **35**, *36*
Common Logic, 146, 165
Conceptual data model, **50**, 50
    dance, 73, *75*
    temporal, 57–59
Conceptual model, 7
Conceptual Schema Design Procedure
    dance model, *See also* Object Role Modeling, 73–75
Constraint, 50
    antisymmetry, 121
    asymmetry, 122
    cardinality, *55*, 68
    complete, 90
    disjoint, 75, 90
    irreflexive, 64, 123
    reflexive, 121
    temporal, 58
    transitive, 86, 121
COVID-19, *26*

D
Dalziell, A., 42, 73
Dance
    salsa, 107
    vernacular, 135
DanceOWL, 107
Database, 52, 66
    integration, 94
    lyrebird dance data, 73
Deduction, 89, *90*
Deep learning, 6
De Morgan's laws, 119
Description Logics, 85, 85, 164
Descriptive Ontology for Linguistic and Cognitive Engineering (DOLCE), 105, *106*
Dirty War Index, 159
Distributed Ontology, Model, and Specification Language, 162, 165
DOLCE, *see* Descriptive Ontology for Linguistic and Cognitive Engineering (DOLCE)

Domain Specific Language, 37, 50

E
E.coli, 27, 129
EER diagram, *see* Extended entity-relationship (EER) diagram
Endurantism, 164
Enterprise model, 70
Entity-relationship (ER) diagram, **52**, 52–57, 146
  attribute, 53
  constraint, 53
  crow's feet notation, *51*
  entity type, 53
  procedure, 66
  relationship, 53
Enzyme, 28
ER diagram, *see* Entity-relationship (ER) diagram
Ethics, 155
Evolutionary diagrams, *see* Cladogram
Existential quantification, 88
Extended entity-relationship (EER) diagram, 54

F
FaCIL, 65
Felidae, 36
Fermentation, 27–29
  mixed acid, *27*
Fillottrani, P.R, 65, 162
First order predicate logic, *85*, 85, 147, 164

G
Gene Ontology, 31, 94–98, 120
Geographic information system, 60
Granularity, 159

H
Halpin, T., 63, 123
Hanspers, K., 45
Hicks, M.H.R., 160
Human-Computer Interaction, 38

I
IDEF1X notation, 55
IE notation, 55
Institute of IT Professionals South Africa, 156

International Union of Biochemistry and Molecular Biology, 33
International Union of Pure and Applied Chemistry, 33

K
Knowledge graph, 3, 98
Kronsted, C., 135

L
Lactose, *34*
Language model, 2
Lemonade, 126
Lewis structure, 33
Logic
  automated reasoning, *see* Reasoning, automated
  semantics, 86
  syntax, 84
Lyrebird, 42, 73
  dance choreography, *44*

M
Machine learning, 6
Mandatory, 56
Mathematical model, 5
Menura novaehollandiae, *see* Lyrebird
Mereology, 121
  general extensional mereology, 125
  ground mereology, 121
  proper parthood, 122
Methodology, 40, 147
  DiDOn, 104
  lifecycle, 103
  Test-Driven Development, 72
  waterfall, 102
Microbial loop, 29–31
Migrant labour, 151, *153*
Mind map, **14**
  benefit, 17
  central concept, 14
  dance, *20*
  size, 20
Modelling language, 50
  design, *163*, 161–167
  expressiveness, 145
  precision, 145
Model-theoretic semantics, 86

Index

N
Nijssen, S., 63
Novick, L., 35

O
Object Management Group, 60
Object-oriented programming, 61
Object-Role Modeling, 146
Object-Role Modeling (ORM), 62–64
  attribute-free, 64
  Conceptual Schema Design Procedure, 67–69
  elementary fact, 68
  example-based approach, 71
  (pseudo-)natural language, 64
  uniqueness constraint, 68
Object type, 50
OBO, *see* Gene Ontology
Ontological commitments, 164
Ontologies, **82**
  automated reasoning, *see* Reasoning, automated
  competency question, 107, 109
  development, *see* Methodology, waterfall
  formalisation, *see* Logic
  foundational ontology, 105, 109
Ontology
  dance, 133, *135*
  history of, 117
  procedure, 131
Ontology (philosophy), **116**
ORM, *see* Object-Role Modeling (ORM)
OWL, *see* Web Ontology Language (OWL)
OWL 2, *see* Web Ontology Language (OWL) 2
OWL 2 profile, 101

P
Parthood, *see* Mereology
Perdurantism, 164
Philosophy, analytic, 116
Physical model, 5
Plankton, 30
Plato, 4, 117
Porphyry, 4
  tree of, *4*
Portion, 126, **128**
Professional practice, 155
Property
  inheritance, 90
  manipulation, 157–159

Q
Question generation, automatic, 100

R
Reasoning, automated, 88–91
  class subsumption, *90*, *91*, 91
  contradiction, 89
  instance classification, 89
  unsatisfiable, 89
Relational model, 52
Relationship, 50
Reverse engineering, 72
Role, 50, 164
Role chain, 86

S
Semantics, 165
Serialization, 88, 166
SNOMED CT, 157
Spagat, M., 160
Structural formula, 33
Stuff, 126
  mixture, 129
  Stuff Ontology, *128*, 130
  traceability, 129
Subtype, 54, *55*, 90
Syntax, 165

T
Tableau algorithm, 89
Top-down approach, 70, 105
TREND, 57

U
UML, *see* Unified Modeling Language (UML)
Unified Modeling Language, 60–62
  aggregation, 61
  association, 61
  method, 61
Unified Modeling Language (UML), 146
Universal, 7, 118

V
Varzi, A., 125

W
Web Ontology Language (OWL), 95, 98, 146, 166, 167
Willi Hennig Society, 34
Wolstencroft, K., 98

X
Xulu-Gama, N., 151

www.ingramcontent.com/pod-product-compliance
Ingram Content Group UK Ltd.
Pitfield, Milton Keynes, MK11 3LW, UK
UKHW022345140125
453680UK00002B/36